SUPPORT AND RESIST

SUPPORT AND RESIST

STRUCTURAL ENGINEERS AND DESIGN INNOVATION

NINA RAPPAPORT

THE MONACELLI PRESS

First published in the United States of America in 2007
by The Monacelli Press, Inc.
611 Broadway, New York, New York 10012

Copyright © 2007 The Monacelli Press, Inc.
Text © 2007 Nina Rappaport

All rights reserved under the International and Pan-American Copyright Conventions. No part of this book may be reproduced or utilized in any form or by any means, electronic or mechanical, including photocopying, recording, or by any information storage and retrieval system, without permission from The Monacelli Press, Inc.

Library of Congress Cataloging-in-Publication Data

Rappaport, Nina.
 Support and resist : structural engineers and design innovation / Nina Rappaport.
    p. cm.
 ISBN 978-1-58093-187-8
 1. Structural design. I. Title.
TA636.R37 2007
624.1'771—dc22
                              2007018651

Design: Think Studio, NYC

Printed and bound in China

INTRODUCTION | 7

ARUP
13

ATELIER ONE
39

BOLLINGER + GROHMANN
53

BURO HAPPOLD
67

CONZETT BRONZINI
GARTMANN
81

DEWHURST MACFARLANE
AND PARTNERS
95

EXPEDITION ENGINEERING
109

LESLIE E. ROBERTSON
ASSOCIATES
123

GUY NORDENSON
AND ASSOCIATES
137

RFR
155

SASAKI
& PARTNERS
167

SCHLAICH BERGERMANN
UND PARTNER
179

WERNER SOBEK
195

JANE WERNICK
209

NOTES | 222
BIBLIOGRAPHY | 226
ACKNOWLEDGMENTS | 231
CREDITS | 232

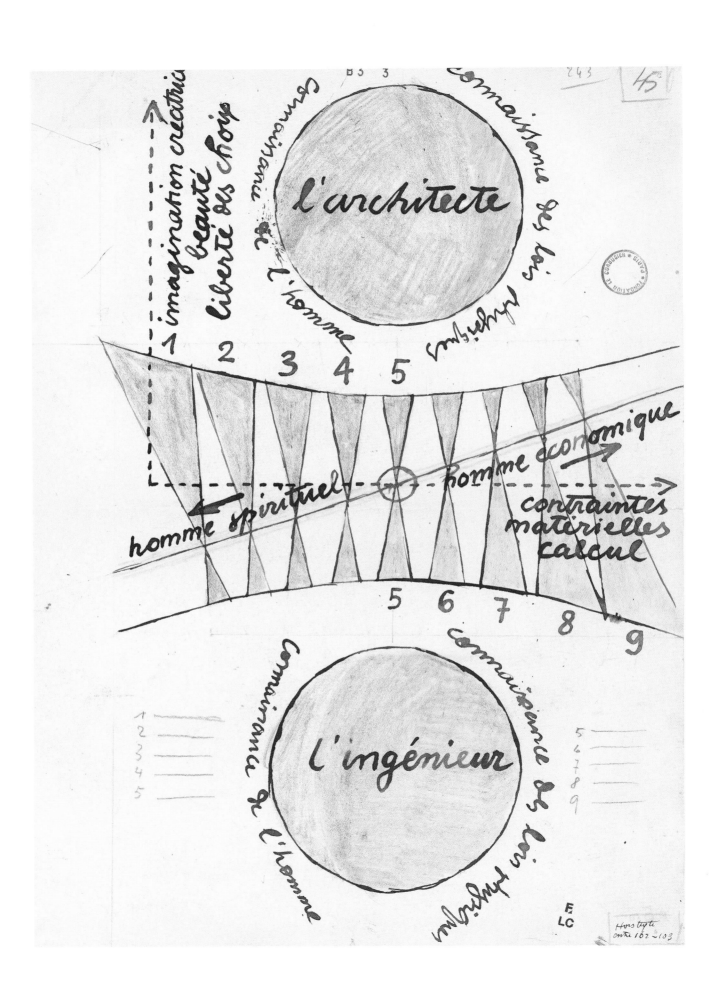

INTRODUCTION

In 1903, a German entrepreneur who made toys in her home decided to expand her business. She asked her husband, an engineer, to design a factory according to the parameters of the time: column-free space for machinery, light and air for comfortable working conditions, and easy location of transport infrastructure. Solving the immediate problems of the program and taking a fresh look at materials of glass and steel, he designed an innovative building with a structural steel framework supporting a double-glazed facade that presaged Walter Gropius and Adolf Meyer's factory for Faguswerke (1911–14). The vernacular project, designed for the Steiff company in Giengen an der Brenz, Germany, exemplified a critical aspect of modern architecture: the anonymous engineer's artistic and technical design innovation. Modern structural engineers who designed factories with strict programs and integrated form, systems, and function expanded the role of engineers.

Today is a parallel time, a moment when the engineer's contribution to the transformation of design and building practice must be acknowledged. With the interest in and complexity of new kinds of shaped-space and the technological advantages of shared computer programs supporting fluid workflow and communication, engineers are playing an essential role in the making of new, complex, and efficient structures. The design engineer's orientation toward formal issues and the architect's collaboration with the engineer from the inception of a project to integrate structure with materials, form, and systems have created a new paradigm. Contemporary practices have formed an exceptional moment in structural design for buildings and infrastructure in a new blurring and overlap of responsibilities.

The French word for engineer is *ingénieur,* which has the same root as the word ingenious. Engineers are inventive, as every solution to a problem is necessarily creative. However, engineering is often understood to be mathematical, empirical, and physical rather than aesthetic, and the process is often too easily reduced to prescriptive results. As Le Corbusier's diagram suggests, the professions are sometimes divergent. Today's innovative engineers are not just calculating stresses and following building codes; they are testing the limits of what is structurally considered feasible with new, unconventional ways of form finding—structures derived

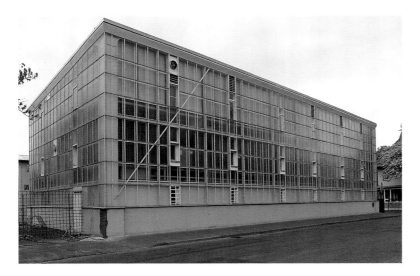

from natural forms, from natural systems, from mathematics—that often lead to changes in codes or building concepts, with unknowns only to be proven in the physical world. As Arnold Pacey, historian of technology, notes, "Although ideas may arise in all sorts of ways that may be described as intuitive or participatory, there is always an obligation to translate them into more rigorous, often mathematical formulations, so that others may understand and check them, and explore their precise implications."[1] Just as Copernicus wrote his scientific treatises in poetic style, so the engineer's pragmatic calculus appears to the eye as an elegant construct, synthesizing all elements. Engineers and architects complete formal and statistical research in the design of a project, playing both artist and scientist.

Engineering is not a science because it is subjective; two engineers will provide different solutions to the same problem, but the structural feasibility of each can be tested. Structures are not only based on empirical study but also evolve from rigorous background and experiential knowledge. Architects and engineers have become partners, each capitalizing on the other's specialization. Creative engineers have expanded the potential for structure, revaluating how structure influences form and the space that it creates, beyond that of "form follows function" to more complex affects of aesthetics. The objectness of a structure, whether visible or invisible, obvious or subtle, can mold or influence the shape of a building. The manipulation of structure, taking it beyond the norm of techniques of the structural support, is where the design engineer's creativity and intuition play a part. The design engineer begins with the site, but the context is within the building itself: the space combined with a knowledge of structural behavior and material potential.

Structures that support architecture, or independent infrastructures, must be designed to work with movement and change. The stability of the structure has to counterbalance and work with the instabilities and unpredictability of nature and with the potential changes of future use. Some engineers use basic principles, rules of thumb, or known and proven systems that have been codified into regulations tied to strict parameters. Others take these rules as a baseline and develop them with performance-based criteria through analysis combined with intuition, just as writers like e. e. cummings or James Joyce invent syntax within an established grammar. Some are close to this edge of divining a structural system through an alchemy of emergent materials and experimental physics, as well as new sciences, but they work within the material properties and constraints of the situation at hand.

This book features the work of fourteen innovative engineering firms and their engineers—Arup, Atelier One, Buro Happold, Bollinger & Grohmann, Conzett Bronzini Gartmann, Dewhurst MacFarlane, Expedition Engineering, Lesile Robertson, Guy Nordenson, RFR, Sasaki and Partners, Schlaich Bergermann und Partner, Werner Sobek, and Jane Wernick. In the projects featured here, the engineer has taken an active design role, questioning what can be reduced or subtracted, what can be connected differently. They might experiment with ways materials bend differently or how glass can become structural, what is the most efficient use of form, what algorithm can be employed to generate form and space, how can the flow of forces be made evident. The aim is not to demonstrate engineering acrobatics, but to show the gestures, moments, details, and connections that enable a new structural paradigm that both influences architecture and has the potential to create new forms and realize the intentions of both the architect and engineer in a new moment of building culture.

Throughout history there have been parallel moments of structural ingenuity. During the late eighteenth century, engineering was distinct from architecture both academically and culturally, but by the end of the nineteenth century, it was often the engineer who made the more inventive architectural projects and prototypes, experimenting with newly available technologies and materials.[2] This development is evident in the introduction of professional courses in schools of architecture[3] and technology, such as the École Central, in Paris, and in the work of Gustave Eiffel, William Le Baron Jenney, or Guillaume-Henri Dufours.[4] Projects in glass and steel became a manifestation of a new industrial and technologically based culture.

As Kenneth Frampton observed, "Any account of modern building culture must acknowledge the crucial role played by structural engineering."[5] Reyner Banham suggested that it was the engineered structures of America that influenced so many Modern architects. European Modern architects such as Erich Mendelsohn and Walter Gropius traveled to America to see the cylindrical grain silos and gridded factories, an industrial vernacular modernism designed and built by engineers, which both justified their functionalist aesthetic and inspired it.[6] Engineers experimented with new reinforced-concrete and prestressed systems for longer spans and curved and parabolic shell structures as seen in the work of early twentieth-century engineers François Hennebique, Eugene Freyssinet, Ernest Ransome, August and Gustave Perret, and Robert Maillart among others. Simultaneously Henry Van de Velde emphasized the engineer's role, "As soon as people know where the true source of formal beauty lies, and who the people who provide it are, they will begin to understand and honor the engineers as they now honor the poets, painters and sculptors and as they honored the architects in the past."[7] Engineers designed the innovative modernist factories while they designed the housing for the machines whose Taylorized assembly line they also configured. Owen Williams (Boots Chemists,1932), Giacomo Mattè-Trucco (Fiat Lingotto, 1926), Pier Luigi Nervi (Burgo Paper Mill, 1962), Felix Candela (Bacardi Rum, 1961), Albert and Moritz Kahn (Ford, 1909), and Marco Zanuso (Olivetti in Brazil, 1954) all integrated systems and structure in their designs.

In 1923 Le Corbusier referred to engineered buildings in *Towards a New Architecture,* asserting that "the Engineer, inspired by the law of Economy and governed by mathematical calculation, puts us in accord with universal law. He achieves harmony." He continued: "Our engineers are healthy and virile, active and useful, balanced and happy in their work . . . and the American engineers overwhelm with their calculations our expiring architecture." Showing American grain silos in their primal geometric forms he exclaimed: "Let us listen to the counsels of American engineers. But let us beware of American architects."[8]

Other engineers designed and built structures. Pier Luigi Nervi was the engineer/architect for the Cinema Augusto in Naples (1926-27) and the Municipal Stadium in Florence (1929), using his new material, ferro-cemento, a fine mesh of steel wire filled and covered by a thin layer of concrete contributing to a new tectonic as the designer of the structural elements. Felix Candela, taught by Spanish engineer Eduardo Torrojo, established a design-build practice in Mexico; Eladio Dieste designed prefabricated concrete and civic infrastructural projects; Heinz Isler continues to design and fabricate concrete shells in Switzerland.

The 1950s was another period of intensive engineering influence on architectural design, when engineers began to collaborate actively with architects. What would the buildings of Tecton be without Ove Arup; James Stirling and Cedric Price without Felix Samuely and Frank Newby; Eero Saarinen and Matthew Nowicki without Fred Severud; Louis I. Kahn without August Komendant; Bruce Graham without Fazlur Kahn; Mies van der Rohe without Myron Goldsmith; or in Italy, Adalberto Libera without Sergio Musmeci? With these architects, engineers integrated structure, materials and form to shape space, synthesizing the rational with the creative.

This book presents diverse engineering approaches of a group of firms that focus on the structural optimization of form, with a commitment to the architect's design aspirations and program in performative rather than prescriptive engineering. The firms are versatile, open-minded and innovative, emerging from a new cultural focus and design-based education with the ability to synthesize form and structure. What is revealed is a series of leitmotifs

ABOVE: Aviary, Zoological Garden, Frank Newby and Cedric Price, London, 1960–65.

OPPOSITE: Steiff factory, Giengen an der Brenz, Germany, 1903.

PAGE 6: Le Corbusier, "L'ingenieur et l'architect" from *La Maison des Hommes*, 1942.

or engineer's preferences in approach based on the use of geometry, natural forms, nonlinearity, asymmetry, broken symmetries, sustainable integration of structures, computer workflow, structural skin, new use of materials or composites, hybrid structures, grid shifts, and what I refer to as "deep decoration" (a structure forming decoration that also fills a space).[9] One can also see the trajectories from engineer to engineer and their far-reaching influences both formally and technologically. But it is necessary to recognize that the architects whose projects integrate structure and design embrace structural exploration. Without that support, the creative engineers could not do this innovative work.

The dynamic between surface, form, and structure has emerged as a theoretical discussion in architecture. Surface has gained attention in its potential for media spectacle or display space, and the creation of effects is influencing spatial experience. Structure by its nature has a capacity to go literally deeper, supporting the new kinds of shaped space, Deleuzian topography, and enveloping spaces that have not been possible. It is perhaps apt that structure has come to the fore, both as structural skin and as a space delineator in an increasingly complex way not only with curvilinear asymmetrical structure, but also with Cartesian grids in dynamic shifts or offset supports. Structure can mold space from the inside out asymmetrically, in nonhierarchical

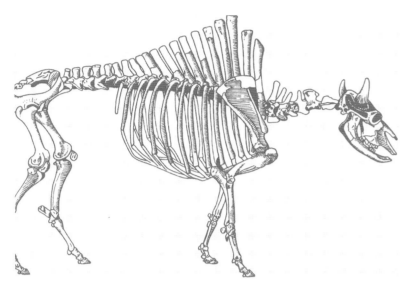 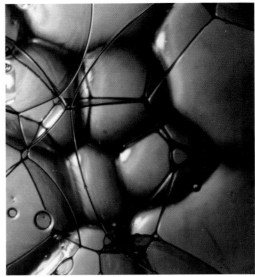

organization related to new discoveries in chaos and complexity theories.[10] Similar concepts of containing space intrigued artists such as Naum Gabo, a Russian Constructivist who studied engineering and investigated a "means of using the void and freeing ourselves from the compact mass." For his organically based sculpture at Abraham Elzas's and Marcel Breuer's De Bijenkorf department store in Rotterdam (1957), Gabo described the design as an "asymmetrical symmetry"[11] similar to the human body. In architecture today asymmetrical and nonlinear forms have resulted in new structural approaches to shaping space. While some might call the forms embodiments of a new structural expressionism, they also enable new kinds of nonhierarchical or non-frontal spatial experiences.[12]

An engineer's philosophy is grounded in the basics of pure math, geometry, algorithms, or generative patterns with the computer. These formulae can develop in an overall network of structural distribution rather than the hierarchical organization or symmetrical proportional systems that evolve into structure that can also be decoration. Some contemporary projects described in this book embody ideas of structure as decoration in particular when the structure becomes a holistic array and cellular in form. Other structural patterns shown here include tessellations that form broken arrays as in the Penrose pinwheel; algorithms that form fractals, which break up and shift a surface; and diagonal grids that get deformed. Computer generated algorithms create a greater potential for the nonlinear and nonhierarchical projects, allowing the form-finding process to be iterative, and enabling a more fluid exchange. Shared computer programs create a common language between engineers and architects, encouraging a closer collaboration than had previously been imagined.

Another foundation of structure and form is an engineer's reference and return to the structure of natural elements. Many engineers have found their structural inspiration in plant life, in a spider's web, a piece of coral, a beehive, soap bubbles, or bone structures. Joseph Paxton was inspired by the giant water lily, the *Victoria Amazonina,* which led to the development of the beam structure for the greenhouse at Chatsworth and the Crystal Palace (1851). With the publication of D'Arcy Thompson's *On Growth and Form* (1917), the analysis of natural structures of plant life and animals gave engineers a greater understanding of form. The links between nature and structure were further drawn in Gyorgy Kepes's *New Landscape* exhibition at MIT in 1951 and in the Vision + Value book series, as well as through microscopic views of nature.[13] More recently, holistic structures in particular, such as crystals, have begun to interest designers. The structure that influences form is as much about its interiority as it fills the space, becoming a part of it, more jungle-like and dense. Thus bubble structures, soap bubbles, interior organization of bones, voids of sponges, and radiolarians have become a fascination for engineers.

Robert Le Ricolais observed that with the structure of bone "if you think about the voids instead of working with the solid elements, the truth appears. The structure is composed of holes, all different in dimension and distribution, but with an unmistakable purpose in their occurrence. So we arrive at an apparently paradoxical conclusion, that the art of structure is how and where to put holes. It's a good concept for building, to build with holes, to show things which are hollow, things which have no weight, which have strength but no weight."[14] Le Ricolais was also inspired by radiolaria as a way to find forms, "forms that encompass the properties of both stressed-skin and triangulated structures. They are just in between: configurations with multiple holes, a perforated membrane in tension working together with a triangulated frame. And this may give an analogy, based on a few topological notions for the arrangement and number of holes, that could bridge the gap between two kinds of structures, and possibly, the two technologies."[15] This understanding of nature and of investigations such as Frei Otto's with soap film for optimization of structures led to experimentation with hybrid tensile structures demonstrating strength in their natural form for

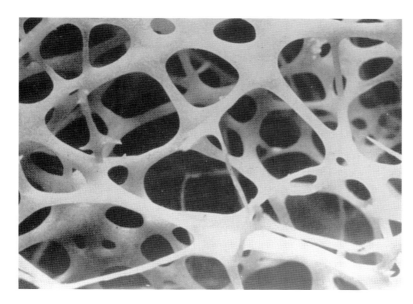 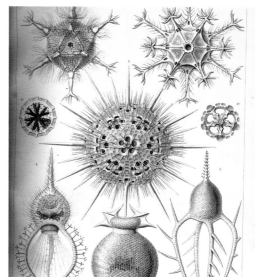

long span roofs. Nature is also used metaphorically for its formal qualities such as biomimicry in tree metaphors and bamboo, as structural supports, which are integrated with environmental systems. Sustainability is considered a holistic element integrated with buildings as engineers harness the sun, wind, and water in projects.

Among the other formal elements that inspire engineers are the potentials for traditional masonry structures, early reinforced concrete, huge steel beams that support exaggerated cantilevers, structural skins made possible by exo-skeletons, and ornamental structural supports. Others are interested in material properties and assemblage of various new materials to create structural form, or emergent materials that behave in new ways opening up structural possibilities. The engineer as a participant in culture also contributes to civic infrastructural projects that might become political or have ramifications beyond the object that they have designed.

This book presents current investigations of structures, inviting a significant voice to be heard through firm profiles, sketches, rarely published construction photographs, and completed projects. The increased expertise with new technologies is now moving to a new phase of collaboration and trajectory of experimentation. Depending on where our cultural interests lie, this could result in new expressions of form and subtle structural configurations. The next generation could take us to new "search spaces" of forms based in nature such as the number of potential arrangements of vertebrae and cellular atomatons, or those structures evolving from genetic algorithms, breeding truss and other construction systems.[16] The structural horizons can be expanded in infinite configurations. The boundaries of the engineering profession have been stretched and the distinction between engineering and architecture blurred. The work goes to physical matter, using geometries in a deeper understanding of space and time, perhaps more pragmatic or liberating, but combined with new physical structural potentials of materials that are stronger and lighter. There is possibility to explore new forms efficiently, optimizing structures, while contributing to new aesthetics. Today, innovative engineering requires collaboration with architects and fabricators, uniting work process with culture and technology beyond what we have known before.

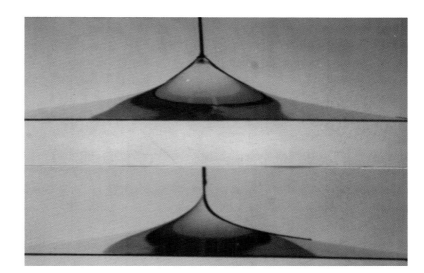

ABOVE: Photomicrograph of bone structure studied by Robert Le Ricolais (top left); Detail from Ernst Haekel, Art Forms in Nature, 1904, Tafel 1: Circogonia. Phaeodaria. Rohrstrahlinge (top right); Frei Otto. Soap bubble film and soap bubble film model, 1960 (bottom).

OPPOSITE: Drawing of skeleton of fossil bison from D'Arcy Thompson, On Growth and Form, 1912 (left); Berenice Abbott, Soap Bubbles, 1946.

ARUP

Sir Ove Arup (1895–1988) founded Ove Arup & Partners, now known as Arup, in 1946 when he was fifty-one and had already amassed a wealth of experience in his own various practices. Arup said that "an engineer must be devoted to architecture. He must have enthusiasm. He must not just be a calculator or analyser, he must be a creative designer."[1] The firm started small, with Ronald Jenkins, Geoffrey Wood, and Andrew Young as partners and a focus on structural engineering and buildings; it has now become a comprehensive engineering practice that encompasses architectural structures, urban design, civil and environmental engineering, acoustics, geotechnics, and virtual computer simulation programs for basically anything that can be built.

Ove Arup, of Danish heritage, was born in Newcastle-upon-Tyne, grew up in Hamburg, and then moved to Denmark to study philosophy at the University of Copenhagen and engineering at the Danish Technical College. He worked with the Danish firm Christiani & Nielsen, first in Hamburg and then in London, primarily on engineering reinforced-concrete buildings, which he had studied at university. A constant inventor, he took out his first patent in 1931 for a cooling-tower design. In the 1930s in London, he began to meet artists and architects during the height of the modern movement. Through modern product design entrepreneur Jack Pritchard (1899–1992) he met Berthold Lubetkin of the firm Tecton and began to collaborate with these modernist architects who were interested in experimental materials and structural expression. In 1938 Arup was part of the Christiani & Nielsen team to engineer the concrete work for Tecton's cylindrical Round House (1932–33) and the ramping Penguin Pool (1934), both at the London Zoo, and for the design of Highpoint Flats (1933–35), an elevated apartment building with load-bearing concrete walls in North London. For Highpoint, Arup designed a revolutionary structural wall and a thin, concrete-slab system to emulate columns and beams, a design based on his earlier work with concrete docks and factories. He also designed a sliding shuttering system that was influential in the "total design" of both the Round

House and Highpoint. Through Tecton, Ove Arup joined the MARS (Modern Architectural Research) Group, an organization of architects looking to improve the architectural and social situations in England. As Jack Zunz said, engineering as a "creative, more positive contribution to the total building process is something that he contributed to engineering in the building industry as a whole."[2]

Finding it difficult to collaborate with construction companies to build in the way he wanted—with forward-looking methods, materials, and technologies—he and a cousin started Arup & Arup, a design and construction firm. Just as Gustav Eiffel, Jean Prouvé, and for a short while Le Corbusier and his brother had done, Arup & Arup became construction entrepreneurs. And although Arup wanted to work closely with architects, he eventually found that Arup the contractor was working at odds with Arup the engineer. In his role as contractor, he would try to keep the price down, sometimes to the detriment of his own innovative ideas. The construction company was not financially successful so in 1946 Ove Arup withdrew from that portion of the business, and a new, strictly engineering firm, Arup & Partners, emerged.

In a lively paper Arup gave as the Maitland Lecture at the Institution of Structural Engineers in 1968, he defined the structural engineer as "one who is competent to design stable and economical structures of different kinds to meet the requirements for which these structures are needed . . . which means that one can be more or less of a structural engineer." The lecture went on to tell the story of a young man named Ernest who was oh, so earnest and who became an engineer through the ups and downs and disillusionment that come with the construction trades, and that his desire to solve the world's problems transferred to "the desire to solve problems that could be solved, problems of designing exciting structures."[3] The story was, of course, Arup's own and described his path to becoming a consulting engineer independent of his original contracting firm, led by his drive to help designers "choose and control" their construction methods and by his keen interest in design itself. He made clear, however, that "we don't build to produce art. We produce useful hardware to fulfill various functions. And we want to get it with the least effort, the least expense. We want value for money, and we can therefore measure efficiency by the simple formula $E = Commodity/Cost$, where commodity stands for what we want to achieve."[4] Previously, in 1948, he had also figured into his equation the special design qualities, or the delight, of a project—those qualities that cannot be assigned a value but that are essential to a project.

The new engineering firm, Arup & Partners, configured structures for numerous projects for architects Maxwell Fry, Basil Spence, and Leslie Martin, among others, in the 1950s. It was a time when architects were enamored with technology as they rigorously and idealistically investigated how to create low-cost projects post–World War II. This was exemplified in the firm's work on the Brynmawr Rubber Factory in Wales (Architects Co-Partnership, 1951; demolished 2001). The innovation in the Brynmawr factory was in the cylindrical shell structure and its nine two-way curved domes, which were supported on V-shaped columns via bow-string trusses. The result was a large, open floor suitable for manufacturing, achieved by the elimination of interfering supports. The shell structure was the largest of its kind when it was built. The Spanish engineer Eduardo Torroja had also developed concrete shell structures, as had the architect-engineer Felix Candela, who was based in Mexico. Following in this structural invention, Arup's experience with reinforced concrete enabled him to assist architects with their design aspirations, but this became more difficult as building codes became standardized. In 1965 the firm expanded and Jack Zunz and Povl Ahm were made partners. Zunz was trained in South Africa and moved to England, first working on Alison and Peter Smithson's Hunstanton School (1954). He returned to South Africa to start up an Arup branch office there. Arup expanded to other countries in a similar way, or by initiating jobs from the London office and then opening an office abroad to be near a project site.

The firm continues to operate on the basic principles that Ove Arup articulated in what is called his "Key Speech" (1970). In this talk to the firm, he discussed concepts of "total design" and "total architecture," synthesized in the collaboration between architects and engineers in design and construction, which he envisioned as one coherent, intertwined process, conceived and carried out together. His philosophy sprung from the ideals of the Modern Movement in the synthesis of art and technology. Walter Gropius had appropriated the concept of total design, or *gesamtkunstwerk*, from nineteenth-century thinkers, and it became one of the basic principles of the Bauhaus as a means to express the combination of art and industry. Gropius used the term in his book *The Scope of Total Architecture* (1937), but the meaning changed over time. Ove Arup adopted it to encapsulate the way the structural process works in integration with architecture. "The term 'Total Architecture' implies that all relevant design decisions have been considered together and have been integrated into a whole by a well-organized team empowered to fix priorities." He also emphasized that "we come up against the fact that a structure is generally a part of a larger unit, and we are frustrated because to strive for quality in only part is almost useless if the whole is undistinguished, unless the structure is large enough to make an impact on its own. We are led to seek overall quality, fitness for purpose, as well as satisfying or significant forms and economy of construction. To this must be added harmony with the surroundings and the overall plan. We are then led to the idea of 'Total Architecture,' in collaboration with other like-minded firms or, still better, on our own."[5] This idea of total design remains a part of the overarching philosophy of the Arup firm and is carried on through their expansion and diversification of their services.

The firm's exploration of structure became part of their broad signature: their early work with concrete, both shell and prestressed; their work with lightweight structures in prestressed membranes, the structural expression of exposed structures in their "high tech" buildings; their new uses of glass to achieve increased transparency; their form finding to make curvilinear shapes; and their high-rise building projects. But more than technical achievement or structural gymnastics, Arup's focus has been holistic, efficient engineering: "All members of the team subscribe to this aim that they all want to help to produce good architecture . . . architecture in depth, so to speak—not just artificially imposed formalism or applied make up—as well as efficient function and economy."[6] As Jack Zunz expressed, "Underlying our work is a freedom or nondoctrinaire attitude toward the problem-solving process . . . Architects need to have some artistic guidance."[7]

Lightweight structures—including tents, air structures, and cables, or anything lighter than the norm of most building materials—became the focus of a division at Arup called Structures 3. Ted Happold, Peter Rice, and other members of this team completed projects most often with Frei Otto, whose soap-film models were used to discern the continuous fabric or membrane surface. Each film is supported at the edges and points, which create the structure in the form of doubly curved, prestressed surfaces. The physical models were able to demonstrate significant error, and computer programs allowed the topology of the net to be analyzed abstractly and predicted how and where the net would hypothetically fall.[8] Alistair Day at Arup & Partners developed the Dynamic Relaxation mathematical technique that could trace, calculate, and analyze the method in which nonlinear, large-scale displacement structures work. The technique resulted in a computer software program, based on Antonio Gaudí's well-known hanging-chain models. Computer simulation became essential to calculating the form, because a net forms its own shape and because the structure and the surface can be integrated into a complete whole, without the technological infrastructure running through the membrane. Arup continued to develop DR computer software, using it to incorporate bracing and structural elements such as ribs and frames, as in the roof of the Schlumberger Cambridge Research Center (Hopkins & Partners, 1985; Ted Happold, engineer, and Frei Otto, consultant), which was suspended from a steel superstructure. The 1988 Hong Kong Park Aviary's four stressed double-curved cable-net surfaces anchored at the ground were analyzed with a similar computer program, Fablon, a nonlinear space-frame program, and Fabcab, a nonlinear tension program.

The Sydney Opera House (1957–73) could, perhaps, be considered the firm's most controversial and public commission at the time. The firm was working with Danish architect Jørn Utzon, with whom Ove Arup felt a camaraderie on a personal level and whose proposal for the structurally determined shell-shaped roofs interested Arup as an engineering challenge. The firm worked on the project for more than six years with Peter Rice and subsequently with Ian McKenzie and Mike Elphick in Sydney for three years. Ove Arup said

ABOVE: Sydney Opera House, Jorn Utzon, Sydney, Australia, 1957–73. Approach from south (left); tile roofing detail (right).

OPPOSITE: Penguin Pool, London Zoo, Tecton Architects, London, 1933–34.

PAGE 12: Pompidou Center, Piano & Rogers, Paris, 1977.

ABOVE AND OPPOSITE: Pompidou Center, Piano & Rogers, Paris, 1977.

of the experience, "If we had known at that time what we let ourselves in for, we might well have hesitated. We underestimated the effect of the scale of the structure." He continued, "The roofs were considered shell structures, as four pairs of triangular shells, but they had to be considered in longitudinal stability as a whole system of four pairs of shells as one . . . It is one of those not infrequent cases where the best architectural form and the best structural form are not the same."[9] It was then resolved that the shells could be shaped from a single sphere, in a more straightforward geometry, so that the precast concrete ribs could be similar length but set at different angles. Other structural achievements were the column-free, concrete folded-slab podium deck elevated on 41- and 56-meter-long prestressed beams; the glass walls that fan outwards; and the mullion system of three U-shaped pieces that follow the complex curvature of the shells and the podium. Arup continued on the project, even though Utzon resigned, until 1973.

In 1971 Structures 3, led by Ted Happold, asked Renzo Piano, Richard Rogers, and Su Rogers to enter the Pompidou Center competition. Unexpectedly, they won the commission, and their dynamic cultural center has proven, as has the Sydney Opera House, to be an extraordinary public attraction. Peter Rice explained the engineers' role: "Once the architectural idea, the large open steel framework, had started to gel, our job, in one page, was to design the framework."[10] As Bryan Appleyard noted, "Rice was sensitive to what the architects were trying to do; he did not, for example, try to talk them out of the 150-foot span in spite of the huge problems it created in the design of the steel trusses."[11]

The building is composed of a framework of fourteen porticoes and thirteen bays and employed 10-ton gerberettes (a cast steel beam hanger named after the nineteenth-century German engineer Heinrich Gerber) developed by Peter Rice.[12] The guts of the building were exposed—the mechanical and ductwork revealed and escalators and walkways placed on the exterior—producing a new honesty of expression. But it also enabled a new type of building appropriate to the contemporary activities inside: an unprecedented flexible exhibition hall in the tradition of world's fairs or the Crystal Palace, but as a reconceived Fun Palace (Joan Littlewood/Cedric Price, 1960–61). The Pompidou Center, as a building expression came to be known as "high-tech" and became a trademark of the work of Arup in the next decade.

Following completion of the Pompidou Center, Arup's office worked with Rogers, Foster, and Hopkins on exposed structures in brightly painted steel. The buildings were organized from modular and structural events, and the technology of the materials is displayed as an aesthetic of joints, details, nodes, and elements. But it is also a type of structural exhibitionism that goes beyond the Modernist dogma of "form follows function," and was conceived of as an expression of technology as a method or approach rather than a stylistic category.

The first of these projects were factories, which seemed to suit the "high-tech" aesthetic, such as the Fleetguard factory in Quimper, France (Rogers & Foster, 1981); the PA Technology in Princeton, New Jersey, (Richard Rogers Partnership, 1984); and the Renault Distribution Centre in Swindon (Foster + Partners, 1982). In all of these buildings, large masts with suspension cables enable the open floors necessary for manufacturing, creating structures like large bridges with the potential for flexibility and expansion of a corporation's facilities.

Many of the firm's projects contain innovative aspects that influence how spaces function and are shaped. The Menil Collection (1981–86) in Houston, a Peter Rice and Renzo Piano collaboration, has an innovative louver system that was incorporated into the lower membrane of a truss. For the Kansai International Airport (1988–95)—one of Rice's last projects with Piano (Rice died in 1992)—asymmetrical trusses, tubular steel ribs, and supports visible in the main public spaces characterized the project. In the Lloyds Building in London (Richard Rogers, 1986) and the Hong Kong and Shanghai Bank (Norman Foster, 1986) the structure was instrumental in shaping open office spaces.

Arup emphasized that "it is the artist's sensitivity, his humanity of personality which speaks to us, if we are receptive to it. But artists differ even more than structural engineers, and there are many ways of producing good architecture. And this is how we wish it to be—we would soon get tired of uniformity." He went on to say, "It is a fact that today the artistic, functional and technical unity must be created by the design. The design record is the construction forethought, which must precede any action in a complex situation. Nothing can be left to chance. We cannot rely on creative craftsmanship guided by a universally accepted architectural idiom and a settled way of life; it is the responsibility of the designers to create harmony out of a chaotic material."[13]

The firm has demonstrated its belief in teamwork and freedom of thought and sees each of its divisions as a community of engineers who can bring their distinct viewpoint to bear on the collaborative design process. The culture is based on experiment and learning, and the firm's broad reach has made it a teaching organization as well. Many prominent engineers, including Ted Happold and Peter Rice, and more recently Guy Nordenson, Chris Wise, and Jane Wernick, have formed their own offices after working for Arup, and they all cite Ove Arup as an inspiration. Arup's approach to the work was a humanist one, and he was seen as having "contributed to the search for a new balance by showing us how to find, or at least recognize, the human face of technology in our design."[14]

Many of the next generation working at Arup have begun making their own contributions to the engineering profession and culture. Among these are Cecil Balmond, Tristram Carfrae, and Markus Schulte. Balmond, now a voice in the academic and engineering world, has expressed his methods and ideas in his books *Number 9: The Search for the Sigma Code* and *Informal*. Carfrae and Schulte exemplify the new breed of creative engineer-designer.

Cecil Balmond, former deputy chairman of Arup, approaches problems of structural design systematically, but also relies on what he calls the "informal" when developing asymmetrical forms. To him, routine engineering concerns itself with Cartesian frameworks; but Balmond moves from orthogonal forms to those based on a rigorous, algorithmic, organization of structure. From this basis in nonlinear philosophy and new theories of science, he focuses on the design possibilities for engineering—not just structure as structure, but in which stability does not necessarily mean equilibrium and is instead closer to nonlinear scientific methods and complexity theories. Within the office, he founded an Advanced Geometry Unit with Charles Walker, as a research-oriented division of Arup where they experiment with new techniques in modeling software as well as creating building forms. "By doing what I do as a designer, structure has moved into a complex interdependency and is not viewed as a dumb skeleton . . . When you break from symmetry, instability is threatening, but it gives its own sense of order, and that is what is so hard for architects to understand: the edge of stability."[15] Making asymmetrical or nonlinear

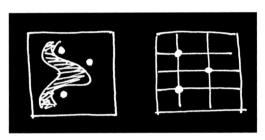

ABOVE: Kunsthal, Rem Koolhaas, Rotterdam, The Netherlands, 1992. Sketches of shifted grid (above); gallery (above, right).

OPPOSITE: Bordeaux House, Rem Koolhaas, France, 1998.

structures, and their accompanying spatial effects, come to fruition involves an in-depth understanding of material capabilities and stress distributions.

Balmond, originally from Sri Lanka, attended Imperial College in London and then went to work at Arup's London office. He has completed the structural engineering for buildings by architects such as Rem Koolhaas, Daniel Libeskind, Ben van Berkel, and Alvaro Siza, each of whom has relied on him to develop solutions for complex designs that they would otherwise not be able to build. His approach is to "encourage free shape, but there has to be a rigorous principle that would create a better solution. I work from an inside process outward. The forms are mathematically based through algorithms based on simple systems and rules." The numerical system is then translated into a three-dimensional structure and is often based on abstracted patterns, knotted or fluid spaces, and networks.

Balmond advocates the "informal" a more relaxed way of thinking about structure: "I don't have to do a structural cage with the column and the beam in a regular skeleton, each time I work on a project," he says. "There is nothing wrong with that, but there is nothing new with it either; it is a basic starting point. But I then ask: Why does a column always have to be straight? Why can't it lean? Why not jump a space and column? Why not skip a beat and create a staccato support system, as in jazz, with a rhythmic idea that leads to architecture. When I work with Libeskind or Koolhaas, we have a dialogue in which we can create a new idea. The theory of the informal is the attempt to put rigor against something that is informal; I am looking for an internal rigor to the structure of the theory."

For example, the Kunsthal in Rotterdam (Rem Koolhaas, 1992), a 3,000-square-meter center for contemporary art, has a simple geometric form that is both logical and out of the norm. The building responds to the site, which drops by 5.5 meters, by transitioning to three levels with ramps and canted steel stanchions. Balmond devised a horizontal-truss bracing system for the roof of the main upper gallery, and takes the form of a sweeping arc in a thin red line that penetrates the roof beams. His intuitive approach is evident in the curve that runs through the span of the gallery space, which is formed from an atypical structure. Balmond started with a standard approach but rather than divide the room equally in a column grid, Balmond devised the idea of syncopated columns, which, as Koolhaas has said, totally liberated the room. Balmond explained, "The solution is simple, but it is not often done. I call it a 'slipped column.' The contractor could execute it and the cost was not much more than the regular column grid." In the lecture hall, Balmond placed the column at an angle. "We were constantly looking for new ways of expressing structure in collaboration with Koolhaas so these are structural statements that can influence architecture, which is what creates a successful collaboration," said Balmond. The slabs are flat slabs, part of which act as "beam strips" connecting with the columns to create the movement frame; columns are placed to allow for a flexible theater space and maximum stability.

In the Bordeaux House (1998), Koolhaas wanted to elevate a box in the air and "make it fly." Koolhaas and Balmond began the design process together by sketching a box on a napkin. Balmond began to draw a line and a box up in the air and started drawing columns to devise a system in which all the columns were tucked under the building and the earth was built up to the house to support one side. If a table is the formal thing, and in the table the legs support the horizontal mass, then he wanted to see what would happen if he moved the legs

the computer calculations to prove the point." Balmond uses the computer as a design tool, not just as a technical tool, and often uses the programs Lusas for Finite Element analysis and Rhino for modeling, as well as the firm's in-house engineering program.

For the Victoria & Albert Museum in London, an unbuilt scheme with Daniel Libeskind (1996), Balmond used a fractal system discovered by the American mathematician Robert Ammann. With the fractal, which Balmond here calls the "fractile," he developed an algorithm for the pattern of tile to cover the spiral shape: three tiles made up the initial base pattern. Each shape is contained in the other two and repeated, and the pattern also builds on the geometry of the golden section. To create a visible pattern, they could emboss the fractal shape on the tile so the light catches

apart and out from under the table. He pulled two legs to one side, with one slipped outside and one still underneath the table. These two extreme moves suddenly liberated the box. "It was so simple and informal or nontraditional a move, like a Karate move with the arm going in and out; here the legs slipped both ways and one jumped up to the top. I broke with symmetries. In the idea of the informal, I rejected the concept of a border, which sets up a closed system. I flipped this so that if you are inside traveling out, the margin is what travels, but there is no border, just like the universe grows to follow itself. I used these notions as metaphors and kept the work informing me. I had no precedent."

While Balmond calls much of what he engineers "informal," he, of course, completes strict computations for engineering analysis using Finite Element (FE) analysis and then he refines the design. Balmond explains that "at first it is like flying a kite of speculation with an informal nature and then I find an internal rigor. Probing it and sketching it and showing the architect what can happen. I then do hand calculations in a very strict method without the computer to get my answers to within five or ten percent, and then I go to

the tiles, defining the shapes. The tiles run around in huge geometric planes 120 feet by 60 feet across and link up together. Although not a new idea, the tiles as applied to a building in that patterning was unique and became a new way to generate a decorative surface.

Balmond worked with architect Toyo Ito on the 2002 Serpentine Pavilion design, in London, employing a structure based on a circling placement of twisted squares and their primary lines of force. They wrapped down and around the space, crossing and folding back to make angular forms that were, according to an algorithm, filled with solid or glazed surfaces for ceiling, floor, and walls. It was essentially a spiraling square. Flat steel was used for the angles and for the diagonal members, which form a dense structural net. The overall patterning of the shell of the space, in crossing lines and planes, makes the skin and structure one, and is more similar in concept to a traditional load-bearing wall than to systems of separate structure and infill. The pavilion is a physical manifestation of an algorithm: pattern and structure are integrated. As Balmond says, "The design started with a simple line that was repeated, releasing architecture from structure, rather than trapping architecture through the structure." The roof panels vary in size, most weighing between 5 and 10 tons, while the biggest wall panel

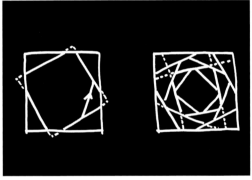

ABOVE: Serpentine Pavilion, Toyo Ito, London, 2003.
OPPOSITE: Sony Center, Helmut Jahn, Potsdamer Platz, Berlin, 2000.

is 20 meters long and 4.5 meters high. Ito and Balmond created the building out of structure, as structure, which produces decoration out of geometry, and complex patterning in a new paradigm.

Also an educator, Balmond was the Eero Saarinen Visiting Professor (1999–2003) at the Yale School of Architecture, where he taught students to use algorithms to devise formal strategies. As the Paul Philippe Cret Professor of Architecture at the University of Pennsylvania beginning in 2004, he initiated a new research study group, the Non-Linear Systems Organization, to explore different ways in which architecture can continue to employ theories from mathematics and sciences.

Balmond is striving toward an architecture with a "new anatomy of organization in a tectonic space. And it is not just about architecture but about systems of organization . . . The radical message to architects, who are used to objecthood and symmetries and externalizations, is that my methods are fundamentally subversive to that. That is not the way of designing buildings anymore . . . So there is a strange area of inhabiting real tectonic space within this unremitting challenging theoretical space." The potential of his spatial work in turn opens new possibilities for architecture and new shapes for space.

Tristram Carfrae is a manager of the Arup Group board of directors, a project leader, and a structural designer based in Sydney. As with so many engineers at Arup, he was heavily influenced by Peter Rice, and was briefly his right-

hand man during the collaboration with Renzo Piano. Carfrae explains that "Peter's great contribution to design is exploration, and exploring what we can do on a project; breaking new ground with better things, not new for new, but better things." Rice introduced him to the idea of humanity in engineering and architecture: "Whatever it is you design, it will be seen by everyone else after, and evoke things, and it is not just something for the glossy magazines. It is the details and textures and quality of human scale, and craftsmanship."

Markus Schulte, from Germany, had joined Arup's New York office but was immediately put on the team for Helmut Jahn's Sony's European Headquarters on the western side of Potsdamer Platz, in Berlin. He returned to Germany and was the on-site engineer once the construction started. Arup was the structural engineer for four of the projects on the Platz, but the centerpiece of their work was the roof over the main plaza, the Forum, at the Sony Center (1992–2000). The design concept, developed with Schulte in a team with Ross Clarke, Bruce Danziger, Mahadev Raman, and Steve Walker, was for an umbrella-like enclosure to be open year-round, built in the shape of a tilted hyperbolic cone to fit over the Forum's elliptical space. The roof is composed of radiating panels of alternating transparent Teflon-coated fiberglass and glass supported by cables that stretch from a compression ring at the circular opening at the top to a space-frame truss, forming an elliptical rim at the base. Suspended in the center is the ring post in the form of a slender, inverted cone. The circular opening results from the mathematical concept of the axisymmetric surface necessary to build the roof, and it employs a ring beam placed along its edge and a kingpost truss along the axis of the cone. The whole structure formed a closed system like a bicycle wheel. The structure of the fabric was derived from the form-finding process that determined the stresses, which are not part of standard building codes but are calculated with each individual project. The glass panels float above the structure on an elastomeric buffer, allowing for movement.

PEDRO & INÊS BRIDGE
COIMBRA, PORTUGAL

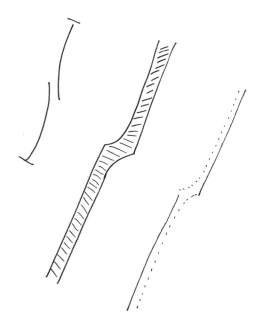

ABOVE: Sketches of shifted center of bridge.
BELOW: Construction sequence in plan and section.
OPPOSITE: Completed bridge.

Working with the engineers António Adão da Fonseca of AFAssociados, Cecil Balmond of Arup designed a pedestrian bridge spanning the River Mondego in Coimbra, Portugal, downstream from the city and flanked by a city park. The challenge was to design a bridge high enough for recreational boats requiring a seven-meter clearance and wide enough for rowing boats to pass through an open span.

Balmond, who had never designed a bridge, was inspired by the site and by the idea of space unfolding. Departing from the straight-line bridge, he imagined it as the arc form of stones skipping across water. In early sketches, he drew two arcs on plan that never met, but passed each other. However, in reality the bridge had to meet in order to cross the river, so he sketched an off-center point at which the two sides joined. In relating the spatial idea to the structure, he saw it as a horizontal bending-moment diagram, resulting in a bridge that went straight, kinked, and continued on.

With that conceptual direction, Balmond devised an 8-by-10-meter, zigzag resting place in the center of the bridge—a transverse shift creating a new spatial experience that reflects the horizontal moment distribution because of the cut and shift of the plane. Shifting the supporting structural line under the deck spine he found that moving it to the extreme position at the edge the bridge would be highly unstable and the whole structure would fall down if the section were to be cut. Thus, the two discontinuous supporting planes shift to the outside edge of the bridge deck over a three-arched steel-box

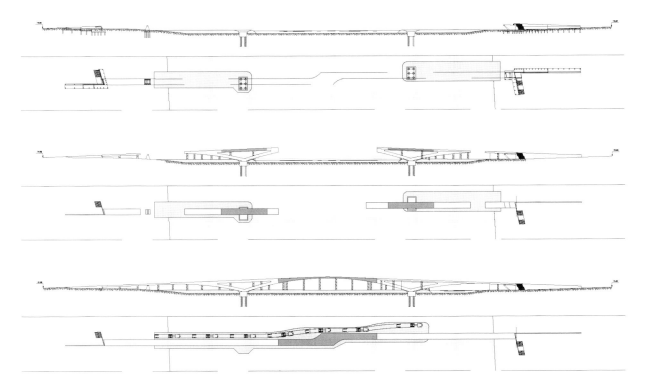

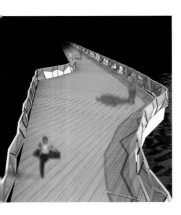

ABOVE: Rendering of bridge footbed.

OPPOSITE: Bridge with point of shifting in distance (above) and at middle edge (below).

girder. Visually from the center, the bridge disappears, turns, and the balance is restored over the whole length. As he explains, "The folded orthogonal to the plane of the deck surface enhances the potentially straightforward bridge with a unique dynamic form on a continuous structure." When the sun hits one side, the beam is highlighted, like a fascia, but when in shadow it disappears.

The concept was similar to his structural strategy for the Bordeaux House: by moving the columns, the box that should have come to the ground is trapped by gravity and is released, and it moves horizontally as a holistic solution rather than an extrusion. Balmond says, "I started with something that never met and the resulting effect was also a mechanical effect in the computer rendering."

Balmond exploits structure beyond what is possible. "I don't believe in the pure form or axial load. The section that takes the axial can take some bending, the material allows you to do something more complex. I always use that trick. There is that classic concept in engineering between torsion and bending moment. There is a saying that one man's torsion is another man's bending. A sophisticated approach can play one against the other."

Balmond designed two inclined ramps supported by two lateral parabolic half arches, forming three arches with continuous triangular frames that support the deck. They meet in the center at a full parabolic transition arch. The asymmetrical arch and deck have complex torsion behavior under vertical loads but have an increased lateral stiffness as compared with symmetrical sections. The bridge deck crosses the river supported by the central arch and is braced by the lateral half arches. In order to control vertical deformations in the lateral northern span, the deck is anchored 6 meters into the abutment.[16]

Balmond desired to create the feeling of a journey in crossing the river, a meandering rather than a straight line. He also was interested in creating the unusual visual effect of the horizon not being visible at the other side of the bridge; it disappears when you turn off center. The low-lying bridge allows pedestrians' feet to be reflected in the water, making a connection with the river. The handrails took on a life of their own as he sketched vines and thicket-like forms. Diagonal planes of the colored-glass panels of the solid balustrades create the effect of movement along the zigzagging handrails he designed to further animate the passage.

The bridge takes its name from a fourteenth-century Portuguese tale. When King Pedro and a Spanish woman, Inês, could not marry because they were of different countries and classes, courtiers murdered her. However, when Pedro became king, he exhumed Inês's body and married her, and the bridge has become a metaphor for this tangential connection.

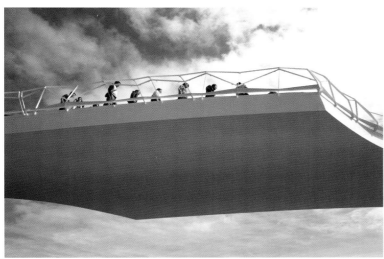

HEADQUARTERS OF THE CCTV
BEIJING, CHINA

The tall building as a type is expanding in concept from the normalcy of extruded vertical towers of stacked floors, core, and window walls to those with loops, kissing towers, hanging balconies, and skyrooms. Rem Koolhaas and Cecil Balmond have explored possibilities for high-rise projects through competitions, such as the Con Edison East River scheme in Manhattan, with Toyo Ito (2001), and a high, slender tower in Seoul for Samsung called Togok (1996), and one in Thailand with slim leaning buildings led to Koolhaas's work on hyperbuildings. The hyperbuildings have frames that slide up and down, and between them, about 244 meters up, Balmond slung a collar with dampers, like shock absorbers, and shifted the tower right through the middle, allowing it to move against the shock absorbers with the connector. It is a loop with a frame and two braces inclined along a tower.

CCTV, China's main television broadcasting company, commissioned a headquarters building by OMA/Rem Koolhaas based on a 2002 international competition. It is intended to be completed as an icon for the 2008 Olympics. The collaborative design process began with a fax sketch from Koolhaas to Balmond of a building as a loop, and Balmond's return sketch added a skin that articulated itself and changed density. As Koolhaas said, "For us it is important to encourage a different kind of engineering, or a different kind of work of engineering, engineering with more imagination, as it is to encourage a different kind of architecture."[17] CCTV had expressed a desire to unite its four organizational divisions into one community, and Koolhaas responded by designing a loop with all related functions distributed throughout the building, but without a central core.

Balmond and Koolhaas began the design process of the loop with a triangulated surface. Arup engineer Rory McGowan analyzed the structure from the stress plots, which as Balmond explained, "came in clumps because you couldn't second-guess this form. The triangulated surface was acting in a strange manifold way; it wasn't an accumulation, it was moving around. And I knew that the skin had to do the work. It wasn't about the cores and a spinal

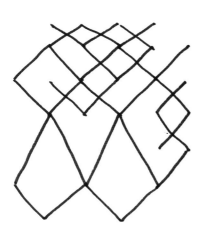 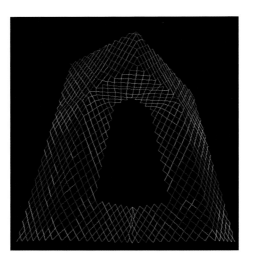 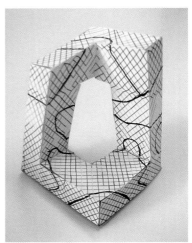

ABOVE: Structural diagram, load diagram, working model used for much of the structural analysis (left to right).
OPPOSITE: Rendering of CCTV and TVCC.

backbone structure, but a skin aided and abetted by cores. We then made a little model in paper about four inches high and marked the stress plots in different colors. And we drew different density crossing lines on it and that was basically the design." That quickly solidified into the generic diamond framework needed for structure. Where the stress was high, they halved the dimension of the frame, so the base 10-meter diamond became 4 or 5 meters, and they doubled the intensity by crossing the diamond. When the structure did not need the intensity, the 10 meters would turn into a 20-meter expanse and encapsulates two or three stories of the building.

For more than a year, the OMA and Balmond teams refined the design with studies that they unfolded like origami and folded up again with numerous variations and color-coded stress levels: red for danger, green for low density, and so on. Thus, the pattern is 90 percent driven by structural efficiencies and the rest by OMA's aesthetic decisions about where to express the structure. The overhang where the two towers meet has the appearance of a cantilever about to fall over. But it is not; it is a continuous, stable loop. The effect is of dynamic movement, and the variable structure expressed in the irregular diagonal grid of the 230-meter tall building makes itself up as it goes, Balmond explains.

But even the idea of movement is in opposition to the concept of what people think should be solid in a structure. The design was at first not accepted in China because of earthquake-safety issues, which includes a fifty- to five-hundred-year probabilities of occurrence. At Tsinghua University in 2003, a government-organized expert review board of engineers were at first nervous. Over nine months, each appropriate government division reviewed the design, from specialists in welding, stress, materials, and environmental systems among others. The design was approved with modifications and ultimately resulted in new standards for seismic analysis in China. The testing revealed that in an earthquake the joints behave plastically so Arup conducted a nonlinear analysis, checking every joint.[18]

Clearing these hurdles the project moved forward. The collaborative design resulted in a continuous rectangular tube composed of two towers, with its diamond-shaped steel frames leaning 6 degrees inward toward each other, supported at the top with an overhanging L-shaped section with a second L-shaped section at the bottom anchoring the towers. New York–based curtain wall specialists Front Inc. refined the diagrid facade pattern by recessing the steel profiles, so that the facade both floats and is integrated with the building as a whole. The circulation system handles 10,000 people in separate elevator groups for each department, as well as corridors across the top of the loop. During the construction, the meeting point of the two beams across the top had to be precise, akin to bridge design, and the air temperature had to be such that there would be minimal to negligible shrinkage of the steel members.

The structure's unusual qualities have put into rational form a new mode of thinking about combining horizontal and vertical structure into a unified, nonlinear dependent loop.

ABOVE: Model of nonlinear shifts in diagrid system.

OPPOSITE: Structural model showing interior structure and core (left); CCTV and TVCC under construction (right).

INSTITUTE OF CONTEMPORARY ART
BOSTON, MASSACHUSETTS

In 2000 Markus Schulte of Arup, began working on a design competition with architects Diller Scofidio + Renfro for Boston's Institute of Contemporary Art (ICA), which they won the following year. The ICA originated in a twenty-first-century spirit, with the question, What is the new museum architecture to be? The institution desired a connection between the waterfront site, on Fan Pier, and their program for gallery spaces. But Schulte said there was a conundrum of wanting magical numbers in order to obtain the right square footage (65,000 square feet) at the right cost. Numerous restrictions molded the project: a maximum building height of 75 feet kept the project from triggering code restrictions for high-rise buildings; the high water table prevented them from digging down; and the ICA desired at least one open-plan gallery.

Diller Scofidio + Renfro, with the engineers, made more than thirty models, and each partner had a voice in the process, in which Schulte also participated. The resulting solution was a four-story building with the main rectangular

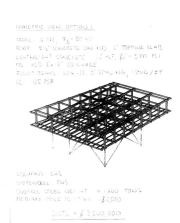

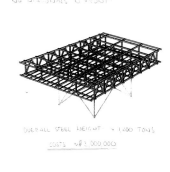
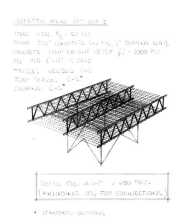
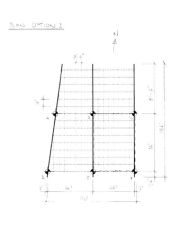
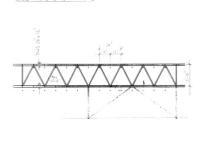

ABOVE: Sketches of structural cantilever options.

OPPOSITE: Upper level theater and entrance.

ABOVE: Steel construction showing middle truss (left) and cantilevered truss (right).

OPPOSITE: West facade.

volume projecting out with an 80-foot cantilever over the site that also created a public space underneath the overhang that is at once inside and outside. Large open spaces on the uppermost and fourth floor—similar in size, Schulte notes, to "a double symmetrical aircraft carrier" are supported by megatrusses. Columns support the megatrusses, and the staircase and the elevator core are placed centrally in the space. Within the massive column-free space, is an east–west gallery with an over 15-foot ceiling height created by the dramatic cantilever. A media theater is suspended below, with seating angled towards the harbor views, fully glazed it seems precariously placed, as though it is dipping through the structure. The Founders Gallery, a long and narrow gallery, spans the northern end and connects the east–west gallery space on the top level creating a total of 18,000 square feet of galleries. A 350-seat theater that seems to be an extension of the exterior walkway is fully glazed, but levels of transparency can be controlled as needed for different events. The building is clad with vertical planes of transparent and translucent glass and metal in a taut skin belying the tough structure.

ICA illustrates how structures create design opportunities, such as for daylight filtering systems, different aggregations of space, and even for new building functions. Schulte was an integral part of

the initial brainstorming dialogues, in a sort of divining process until the most powerful and integrated solution was selected. Some engineers use the rhetoric of design and the designer's language, but Schulte thinks that diminishes the true design capacity of engineers, who do indeed design, but differently from architects. Schulte says that when structure becomes a generator of architecture, the rhetoric is not important, engineering is. The solutions come out of the energy and the spirit at the table. And Schulte emphasizes that, "material and technical innovation is meaningless unless paved with intellectual process, powerful ideas, and open opportunities."

WATER CUBE
BEIJING, CHINA

Arup is the engineer for the Beijing National Aquatics Center, known as the Water Cube, for the 2008 Olympics with PTW Architects of Australia, and the China State Construction and Engineering Corporation (CSCEC) and the Shenzhen Design Institute (SDI). PTW's scheme, designed for the 2003 competition, features five pools enclosed within box walls that are like blue bubble-pack. This project exemplifies the collaboration of the engineer and the architect but also demonstrates the engineer's ability to play a defining role. Tristram Carfrae—chairman and design and technical director of Arup's Sydney office, and who also teaches at the University of Sydney and the School of Built Environment, University of New South Wales—was the coordinating engineer for the project. He emphasizes that the Water Cube deviates from the typical structural problem of achieving an architectural form and the services required to support it from within. With the Water Cube, Carfrae focused on how structure itself fills space, a process closer to that of form making in semi-solid natural objects such as beehives and multiplying cellular crystals.

On the day of the announcement for the Aquatics Center competition, Carfrae organized the Arup Sydney team's six disciplines, and together they decided what from each technical perspective would be important to present to the architects before the architects even picked up their pencils. A four-week period followed in which representatives from each division struggled independently for an idea. During this period, an addendum to the brief was issued, saying that the southern end of the center must be public space.[19] The team was then asked to include a water polo facility, an ice hockey rink, clubs, and bars to generate revenue, and they decided that a square would more easily accommodate the additional facilities, and thus the box became the solution, sited adjacent to the circular stadium designed by Herzog & de Meuron (also with Arup as engineers). The square signifies the Earth and the circle, the heavens, in a harmonious yin and yang relationship in Chinese culture.

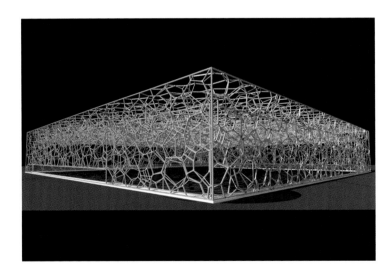

To enclose the space and create the architect's desired bubble effect, Carfrae considered using vertical tubes that would appear as circles on the outside of the building and would allow light to filter in rather than glare. But he wondered, "What would make the wall horizontal? What would the joint between them be? And how would it transition from the vertical to horizontal?" He recognized that the geometry appeared to be an organic structure, as structure in space. "Usually we are interested in planes. Bucky Fuller had structure covering surfaces. My question was, What about it occupying space?" He researched the classic question of bubble theory, referring to Frei Otto's investigations into soap bubbles, because he was inspired by the shapes of cell formations and soap-bubble arrays and how their connectivity rather than their structural form divides space. Soap bubbles will naturally dispose themselves in a series of equal volumes so that the partitioning area is minimized. Carfrae researched the work of eighteenth-century French mathematician Joseph Plateau, who had observed that soap bubbles agglomerated with the walls coming together in four lines to form a

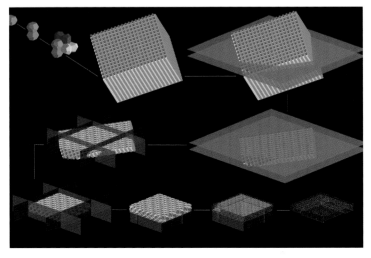

ABOVE: Rendering of exterior showing ETFE bubble facade.

OPPOSITE: Rendering of facade framework (above); computer modeling of volume and facade (below).

tetrahedron, though he never went on for further proof. Plateau's answer was highly regular and the question of how to structure space was only partially answered by the nineteenth-century physicist William Thompson Kelvin, who also was investigating the division of three-dimensional space.

Carfrae was also made aware of the 1993 studies done by Denis Weaire and Robert Phelan, physics professors at Trinity College, Dublin, which analyzed the connections between bubbles. Weaire and Phelan created solids with fourteen faces (two hexagonal and twelve pentagonal) and made multi-sided bubble shapes, which, when they were combined with dodecahedrons, fit together. The space-filling concept of foam fill could be buildable and repetitive, but looked random, just as in the space-filling patterns of cells and crystals. Carfrae plugged it into a CAD program and instead of orienting it on a Cartesian grid skewed it at an angle, integrating it with the architectural ambition of the project. The structure would combine the surface pattern with the internal structure, creating organic patterns that are both functional and structural that results in the concept of "deep decoration."[20]

Numerous structural issues had to be solved once the main form was determined, such as how to keep the structure lightweight enough for the long-span roof and how to meet the seismic requirements. The ductile space frame generated from the geometry would perform effectively, but Carfrae worked to make the structure even more efficient: "It is a leap for me, it is driven more by a big idea than specific structure."

The resulting structure is a 177-by-177-by-31-meter Vierendeel-type space frame based on a geometric cell of twelve pentagons and two hexagons that can be infinitely repeated as a network. Composed of a series of 22,000 steel tubes welded to 12,000 spherical steel nodes, the structure is varied according to loads, minimizing the steel tonnage and satisfying the design requirements for the long-span roof structure. Using an iterative process made possible by a structural optimization program, the engineers defined the element sizes and analyzed 57 million design constraints for weight and strength, adjusting the node elements and tubes to each new move. Computer programs simulated and analyzed fire egress, crowd circulation, and environmental systems. The vast, greenhouse-like design system makes the building highly energy efficient, harnessing sunlight, which penetrates the Ethylenetetrafluoroethylene (ETFE) cladding, to heat the pool.

The ETFE cladding, foil cushions in a lightweight recyclable material that are also self-cleaning and durable, results in an array of 4,000 bubbles—some of them 7.5 meters in width—defining the cube. Although the repeating cells are of only two different sizes, they are randomly distributed and are cut arbitrarily by the building's surface planes, creating a nonrepetitive pattern. The material visually refers to cellular forms and soap bubbles. Functional in an efficient structure, the cladding exhibits qualities that are also decorative, resulting in an innovative space. The roof is composed of bubble segments with pattern borders on both sides, which form the separations between the bubbles as well as the structural force. The firm used Rapid Prototyping to physically model the structure with sintered nylon powder and solidified epoxy resin in a stereolithographic method. But the model was larger than the prototyping machine, so it was made in four parts and assembled later. The transparent model cladding was handmade in China from the same three-dimensional CAD drawings, and the pieces fit together perfectly.[21]

As in nature, the mathematical world is expressed in the morphology of this building resulting in an aesthetic of calculations and nonlinear development. The overall effect is in unity with and emerges from the structural system in a complex form.

CLOCKWISE FROM LEFT: Installation of ETFE cladding on facade; completed section of ETFE cladding; interior rendering of swimming pool.

OPPOSITE: Rendering of elevation showing structural filigree.

ATELIER ONE

Neil Thomas and Aran Chadwick founded Atelier One to design structures for art, architecture, and infrastructure with an inventive and fluid approach. Works by the London-based firm range from small installations and stage sets to large-scale urban developments: architect Mark Fisher's robotic sets for U2's POP Mart tour (1998), Zoo TV (1993), and the 2006 Winter Olympics in Turin; rapid-build prefabricated houses such as the White Cube 2 in London (2003); the Garden installation at the Prada Foundation in Milan by Marc Quinn (2000); and the Singapore Arts Center (2004) by Michael Wilford & Partners.

Atelier One combines an ability to see the built environment from new perspectives with a design flexibility in a repertoire engaging nature and art.

When working with artists, Thomas finds a synergy making sculptures structurally competent, which for him "have a completely different feel from architecture. The rules are different; the constraints are different. Artists are usually reaching for an idea; in architecture, the constraints are varied and moving around all the time. The difference between architecture and art projects is that the art projects do not necessarily have to support people, and you can push in different directions; the rules you have to work with are not the same set of rules; you can move into a world that is more open; there is more room to suggest things that you otherwise couldn't do."

One early artist collaboration was Rachel Whiteread's *House* (1993), a nagative cast of a Victorian terrace house. Atelier One proposed a system of spraying the interior of the house with concrete, giving it stability while maintaining the level of details available in plaster and thus contributing to the project as a whole. The contrast forms but does not control the project, just as the structure informs but does not overwhelm the artwork.[1]

Atelier One has worked with Anish Kapoor on a variety of projects such as the installation *Taratantara* at the Baltic Flour Mill in Gateshead, England (1999), the initial concept for *Marsyas* at the Tate Modern in London (2002), and *Cloud Gate* at Millennium Park in Chicago (2005). In Gateshead, Kapoor conceived of

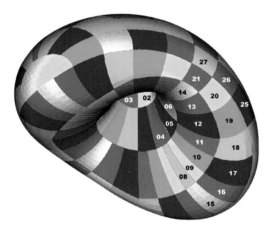

a fifty-meter-long tubular membrane installation. Thomas and Chadwick, understanding his idea of a fluid structure bound in a volume, developed the three-dimensional surface of the membrane structure. The 35-meter tall and 50-meter long trumpet-like form consisted of forty-six strips of fabric high-frequency-welded and tensioned with D-rings, and fastened with bolts to a frame at either end of the interior of the mill. All membrane surfaces have a strange moment when they appear to be doing something that they are not. In the case of Taratantara, the membrane creating a trumpet shape is foreshortened forming an optical illusion of increased space within the industrial framework.

Cloud Gate was commissioned as part of Millennium Park's extensive art and architecture program in 2005. Atelier One devised not only the structure and fabrication for the sculpture but also the transportation and installation of the 22-meter-long steel "bean." The artist's design was scanned into the Rhino computer program so that the engineers could complete a Finite Element analysis model, then transmit it to the fabricators for production. The thin structural shell in polished stainless steel was built in Oakland, California, with a six-axis milling machine typically used by the airline industry. The extremely thin plates with double-curved surfaces are suspended from an internal steel armature similar to that of the Statue of Liberty. Internal welds are concealed, and interior hidden soft connections allow thermal movement between the shell and the armature, while creating the appearance of a monolithic homogenous skin.

With artist Lucy Orta, in 2005 Aran Chadwick designed *Dream Gateway*, an enclosed pedestrian bridge, for the Lee Bank Middleway in Birmingham, England, as part of the area's revitalization. Incorporating concepts of the

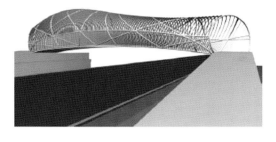
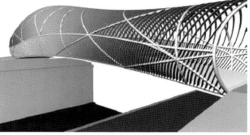
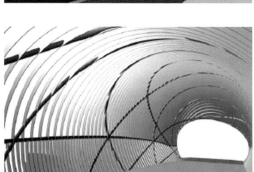

LEFT: Lee Bank Middleway Bridge, with Lucy Orta, Birmingham, England, 2005–7.

OPPOSITE: Rachel Whiteread, *House*, 1993 (top); Structural stress diagram for Anish Kapoor, Cloud Gate, Millennium Park, Chicago, Illinois, 2005 (middle); Anish Kapoor, Cloud Gate, Millennium Park, Chicago, Illinois, 2005 (bottom).

PAGE 38: Anish Kapoor, Taratantara, Baltic Flour Mills, Gateshead, England.

cosmos and organic form, Orta conceived of a bridge that symbolizes a time warp in the pedestrian's journey. Connecting two communities, the bridge will visually vibrate from the light that emanates from the interior. To achieve this effect, Atelier One devised a system of circular steel hoops that frame the bridge with its horizontal contour bands running lengthwise. At critical structural points, the hoops form a truss to support the single-span bridge over the dual roadway. The contour bands, made of denser steel, provide the required balustrade. The bridge, approximately six meters high in the center, slopes down to two meters at each entrance. These "entrances" are themselves a new way of conceptualizing a pedestrian bridge as a room to inhabit rather than as a platform to walk on. The steel bands also contribute to a dynamic moiré effect when the bridge is viewed from the side. The completed bridge will be trucked to the site for installation in a civic as well as a structural event.

In their collaborations with architects, Atelier One emphasizes integrating environmental forces, such as the effects of wind and sunlight on a building, incorporating natural forces rather than fighting them. Instead of hard, stiff buildings that counter gravity, Chadwick and Thomas design structures that are flexible and responsive to nature. Their process is organic in that it uses qualities of nature and geometry as points for design synergy between form and structure. Thomas and Chadwick see engineering as a process in the natural evolution of a structure. Thomas does not suggest that this practical approach alone makes architecture but that it is assimilated in the architectural design: "Architecture is about people using space and then about desire, so that the engineering is apparent from the form."

For the competition proposal for Stadium & Sports Campus Ireland at Abbotstown, Dublin (Behnisch, Behnisch and Partners, 2005), Atelier One merged environmental and structural systems with the building design by ascertaining the impact of wind and sun on spectators. This exploration guided the concept for a series of fractured canopies that would spiral out of and into the earth; simultaneously, the ground would rise up to meet the roof. Deep trusses would shift off-center to provide transparency rather than shade spaces between the dynamic roof plates.

As Thomas explains, the engineers' method is both reductive and expansive: "The expansive idea is that you solve a problem and then move to the next problem and solve that. It becomes a layering of discrete answers to each, one step at a time, to reach an overall solution. The alternative is a reductive approach, considering all the issues and constraints together, to develop an idea that resolves them in a single solution." Larger-scale projects provide opportunities to resolve significant issues with new structural paradigms and to reach the architect's goals in a way that combines both structural and environmental concerns. The firm designs new structures without pretense, engaged in process, so that translating an idea into a form is fluid; there is a "liquid threshold," as Thomas says. "It is not rigid and fixed. That line can approach apparent chaos, as in Federation Square, which demonstrates a fine line between being chaotic, and therefore undoable, and finding the order within, while maintaining the impression of chaos. If the project is both large and totally abstract, it is also likely to be too costly, so it is about finding an order within an appearance of chaos."

A deep engagement with process, whether for an artwork or an architectural construction, liberates Thomas to design many types of projects at many scales, with intuitive sense of a structure's essence rather than just as a mathematical calculation on the computer. As he explains, "Computers have made it possible to generate and analyze extremely complex surfaces. The skill is envisioning a rational structure to create those surfaces in the real world and how that might be engineered and built."

Thomas studied architecture and then engineering at Leeds University. He has a particular interest in how things work and in the nature of materials, and he has been influenced by his father, a planning engineer. At Buro Happold from 1980 to 1986, Thomas worked with senior partner Ian Liddell

on, among other projects, the Welsh National Eisteddfod demountable theater, which they won in a competition. On site he assisted in actually erecting the building, which in turn informed his engineering background. Thomas was inspired by Ted Happold's ability to never say no to the possibilities of a structure and his interest in the "why not" of a project. Thomas then worked with Anthony Hunt Associates, founding the subsidiary Hunt Projects to work at a smaller scale. When Hunt's office was absorbed into a larger company, Thomas left to found Atelier One in 1987 and Aran Chadwick joined in 1992, becoming a director in 1997. Chadwick's previous work included projects such as the Canary Wharf tube station shell structure and post-tensioning remedial supports in historic buildings. The name, Atelier, was chosen to represent their experimental and artistic approach. In 1991, Atelier Ten, an environmental engineering firm, was founded as an offshoot, headed by Patrick Bellew and Steve Marshall in London and now New York. Atelier Ten inspired Atelier One to focus on the environmental influences on structure, allowing structure to accommodate the environment.

Thomas and Chadwick teach engineering as they practice it—as an integrated discipline—at schools such as the Royal College of Art, Bartlett School of Architecture, Architectural Association, and Yale School of Architecture. They start with basic ideas, with the tools and principles of materials and structure: the force of the wind, the live loads of snow, furnishings, and people. "Architectural students must think about engineering as immediately as they think about architecture," says Thomas. "The architecture student is really a design team embodied in an individual. When they think about the envelope and the form, they must consider themselves as engineers, and think of structure and environment."

Thomas also teaches about engineering through examples of failure, commenting that every building is as a prototype. He believes one way to learn is to study past mistakes, as in the Tacoma Narrows Bridge, Millennium Bridge, and Roissy-Charles-de-Gaulle Terminal E.[2] As Thomas says, "If you push structures to the limit, sometimes you come across something that no one has thought about before. But how can you be responsible for that? So there has to be a bit of conservatism in design because there are life and death issues; understanding failure exposes an understanding of structures." Yet Thomas also believes that risk taking, finding and then pushing the limit, is necessary.

In some of these projects at the limit, Atelier One is called upon to make nature itself. Their work with set designer Mark Fisher on performance stages for U2 may be compared to special effects in cinema, experimenting with ways to control nature and chaos to create a phenomenological experience. In addition, they are working on a series of projects with the British artist Marc Quinn, whose work is about life and death. Quinn created the project *Garden* in 2000 as a sort of "hortus conclusus": a garden of infinite beauty and immortality; an environment created by frozen plants from various continents and seasons. The installation, in

a stainless-steel refrigerator (-20°C), consists of flowers frozen in silicone (which stays liquid to -80°C and does not chemically react with the flowers). Thomas has worked with Quinn to make a small rainbow in a sports hall at a school near Quinn's studio, building the mechanisms to create the light and the air, using the laws of physics and nature. Thomas has also experimented with making clouds by controlling the stratification of mists. For The Environmental Angels at Clarke Quay in Singapore (Will Alsop, 2005), Atelier One designed a roof structure over several shopping streets; treelike columns, called "angels," contain environmental controls and modify the exterior air temperature. Also in Singapore for Gardens by the Bay, with Grant Associates, Wilkinson Eyre Architects, and Atelier Ten environmental engineers, Atelier One has designed the engineering for Supertrees, vertical gardens that will be part of an environmental web for a vast interactive garden park project to be completed in 2010 that integrates structure, nature, and sustainable design goals.

Atelier One's projects combine environmental and structural engineering holistically, often invisibly. The influence of nature is not conscious, but as Thomas says, "You can't fail to incorporate it if you follow a particular route. It manifests itself, not as an organic form or an arbitrary shape. If you can, accept that trees are the way they are because of wind and sunlight. We are applying those same rules to find new and natural solutions."

ABOVE: Supertrees in proposed Gardens by the Bay, with Grant Associates, Singapore, 2007–10.

OPPOSITE: U2 Pop Mart Tour, Mark Fisher, 1998 (above); Marc Quinn, *The Overwhelming World of Desire (Paphiopedilum Winston Churchill Hybrid)*, garden installation, Cass Sculpture Foundation, Goodwood, West Sussex, England, 2003 (below).

SINGAPORE CULTURAL CENTER
SINGAPORE

James Stirling Michael Wilford & Associates won the design competition for the Singapore Arts Centre (SAC) Theatres on the Bay project, a theater and concert hall complex called the Esplanade, in 1992. After James Stirling died, the firm changed its name to Michael Wilford & Partners and formed a team for the project with a local firm—DP Architects, led by Koh Seow Chuan.

The original proposal had wrapped the two music hall venues inside a pair of glass shells. The government committees considered the design too Western and alien, however, and asked the designers to revisit the shell forms and cladding. As a result, in 1994 the design team hired the environmental (Atelier Ten) and structural engineers (Atelier One) to analyze how the sun and the prevailing wind direction affected the building. They determined the environmenal orientation and outer shell design. The designers referenced characteristics of indigenous shelters, such as leaf layering of roofs, which provides complete waterproofing, shading, and raised leaf end-points, which allow sunlight into interiors.

In redefining the original shell, the engineers combined a double-layered structure and devised exterior triangular shading panels, oriented to provide shade inside the building while allowing views to the outside, like the leaf-covered shelters. Feng shui analysts prescribed softer building elements, so the team placed the triangular, or half-pyramid, shapes along the vertical sides, pointing down, and all were redesigned with rounded nodes. Where the glazing was horizontal, the triangle was elongated,

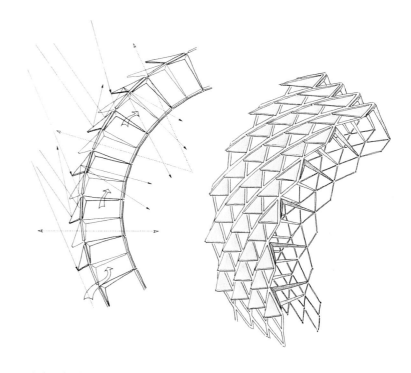

and the shades on the vertical faces, which received the least sun exposure, are smaller.

By permitting support from the acoustic box, a single layer shell structure evolved. Its overall form was defined using software for pneumatic membranes. Then the surface is converted to a compression shell (a twenty-first century hanging chain model). This method allowed the surface to be varied in volume to accommodate the spatial requirements. To this is applied a structural mesh of equal link lengths. As the mesh traverses the surface, the mesh lozenges change to accommodate the curvature, creating an exciting patterning within the structure. The glazing envelope and the shading are supported at the nodes of this structural mesh. By employing the changing shape of

ABOVE: Diagram showing angles of the sun in relationship to shading devices.

RIGHT: Elevation of the exterior skin and sectional drawings of complex form.

OPPOSITE: Theater buildings, view from the west.

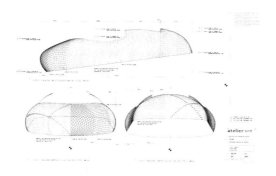
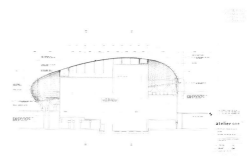

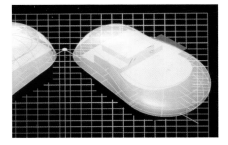 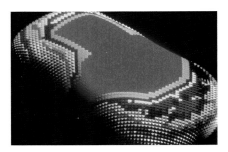 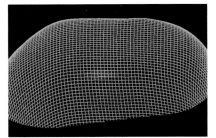

ABOVE: Computer modeling of structure and stress analysis; aerial view from southwest.

OPPOSITE: Mock-up of facade system (above) and detail of shading devices with structural framework (below).

the mesh, the shading is adjusted with the curvature from diamond shapes at the gith curvatures to square at the upper sections where the surface radius is greater, i.e., flatter. The sunshades themselves are rhomboids, folded on the shorter length of the parallelogram, forming a small pyramid. One side is placed against the volume and the other lets light inside. All shades have a constant edge of 1.5 meters in length. This maintains a consistent, fluid form, with only twenty-five different computer numerical control (CNC)–milled templates.

A single layer welded-steel structure of tubes 200 millimeters in section holds a grid of glazing frames. The space frame structure is reminiscent of Buckminster Fuller's dome, but the spatial composition is more refined and elongated. The utilities are placed in interstitial spaces, and can modulate the volume in response to the climate, to optimize air-conditioning.

For the Lyric Theater, a more open arrangement allows views to the sea and the angle shades the sun. At the ends near the theater, the shades are fixed in a lower position. The perpendicular mesh shades on the side of the Concert Hall are organized on a diagonal axis to block the more direct sunlight while allowing for differentiation between the two volumes.

The overall building shapes resemble a local tropical fruit called durian, which is spiked on the outside but smooth on the inside. The goal, however, was not to design a biomorphic form but to expose the structure as nature's architecture, mimicking nature's structures, such as the leaves used to waterproof a roof, but using contemporary materials. Neil Thomas explained, "As the design developed, through the strict use of environmental parameters combined holistically with the structural mechanisms to achieve the architectural desire, the outcome is architecture representative of its climate and cultural aspirations."

FEDERATION SQUARE
MELBOURNE, AUSTRALIA

Federation Square in Melbourne, Australia, was designed by Peter Davidson and Don Bates of Lab Architecture Studio with Bates Smart after they received the commission in a 1997 competition. They completed the project in 2003. The civic development project, which decks over one section of the city's railroad yards, creates a dynamic 7,500-square-meter public space with cultural institutions in the surrounding buildings. Basing the conceptual design on the angular connections of the fragmented city, the architects with Atelier One and Atelier Ten sought a system of triangulated fractals for the building facade from which the design could be generated. The engineers suggested the Penrose pinwheel algorithm as a geometric framework to form the basis of an intricate variegated structure from a limited kit of parts for structure and tiling. The structure's coherence pervades the project—from the tiled skin of local stone and the zinc that reflects the sunlight, through to the structural skeleton. Simultaneously holistic and expansive, the structure becomes decoration when it engages in the act of defining a spatial effect.

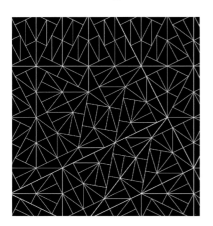

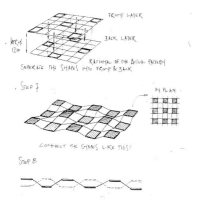

Neil Thomas describes the algorithm: "Membrane structures use an iterative method to fit a series of "variable" triangles to a doubly curved surface. These variable triangles are easily accommodated within the "soft" fabric. The Penrose pinwheel uses the same triangular element throughout and thus the regeneration of each surface requires interaction between surface and boundary to endure that the single module is always maintained." Discovered by Sir Roger Penrose, the Penrose pinwheel comprises a set of aperiodic tiles with fivefold symmetry (impossible in periodic crystals), used to explain the structure of some "quasi-crystal" substances.[3] Previously a fivefold form had not been used as an algorithm, because of the gaps left when tiling a planar surface as it is a broken symmetry. The Penrose pinwheel uses triangles to create continuous patterning, a tessellation that the engineers incorporated into the design for Federation Square. As Thomas notes, "This generative method was then combined with geometric stiffness induced into a single plane surface, much like the effect when you crumple a piece of paper. A slender unstable surface becomes rigid. By rotating the combined triangular elements, the appearance of disorder is obtained, although made from the same modular elements."

The algorithm also suited the architects, who wanted to break out of the norm of walls, windows, and apertures in general; they designed a system where the facades of the main buildings are composed of a rain screen system, with the outer layer of triangular tiles left open jointed. The inner layer is a thin aluminum sheet, which constitutes the weatherproof skin. By using an integrated cladding and structural system to wrap around the building, the architects form and characterize internal spaces by means of the shape of the holistic building envelope. The Penrose algorithm allows for the shape of the envelope to change and fold dynamically, with the plaza surface also being defined by this triangulated system. As the triangle is turned and shifted, the compositional array organizes into a pattern. The pinwheel triangle can be divided into five equal triangles so that they all can continue to be subdivided but stay in proportion to one another, creating an asymmetric fractal system. One set of panels is mounted to a framework, and the framework, becoming in turn a larger panel, is attached to a triangular galvanized steel frame. The pinwheel construction is appropriate for the facade in that no sealants are required, so that it creates a cleaner, smooth surface even where the majority of the triangles meet. Each panel sits on a substructure of the building's skin.

The glazed atrium structure also exemplifies a new relationship between structure and decoration.[4] The atrium is an exposed structural form that in its repetition and distribution proposes ideas of patterning and algorithm as structure. The frames are four- to five-sided irregular polygons connected by in-plane diagonals, with strength

ABOVE: Structural framework at atrium corner (above) and during installation of glass cladding (below).

OPPOSITE: Facade tiling system. Penrose pinwheel patterns (top); sketch from grid to deformed grid (middle); zinc and stone panels (below).

and stiffness provided by truss and moment-resisting actions. The two skins are not parallel and are connected by diagonals that provide stability. The open network of this framework of 3,000 members seems to fold in and out within the open void formed by the outer and inner glazing.

The Penrose works at multiple scales, as the secondary steel structure with a pinwheel-generated frame, a finer interpretation of the Penrose geometry, supports the glazed panels. The structure is visible through the glazing and informs the comprehension of the rest of the structure, and spills over to the panel system sharing the pinwheel non-orthogonal patterning. The new form and surface patterning merge physically in the building, creating new spatial arrays and spatial interiority of new shapes.

CLOCKWISE FROM LEFT: Diagram of structure; wrapping diagram of facade cladding; atrium model; model of structural meshwork.

OPPOSITE: Federation Square completed project.

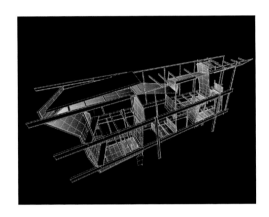
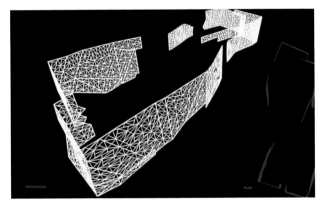
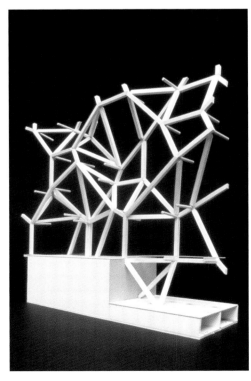
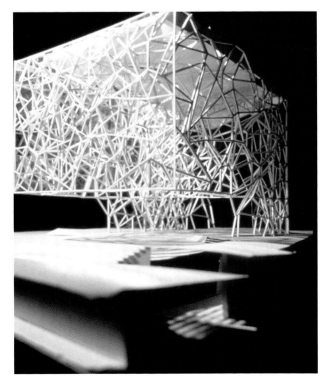

BOLLINGER + GROHMANN

Klaus Bollinger and Manfred Grohmann, who met while studying civil engineering at the Technische Hochschule Darmstadt (Technical University of Darmstadt), founded their firm, Bollinger + Grohmann, in 1983 in Frankfurt. The thread running through the firm's philosophy is the ability to make architects' visions become reality—an "engineering without ideology"[1] that liberates structure and, with it, form. Their adaptability and flexible working methods have contributed to their successful collaborations with over fifty different architects of widely varying design repertoires—architect Peter Cook, Wolf D. Prix, Kazuyo Sejima and Ryue Nishizawa of SANAA, Frank Gehry, Dominique Perrault, Zaha Hadid, Bernhard Franken, and Bolles + Wilson among them. Synergy evolves through communication and sensitivity to complex architectural designs. Peter Cook emphasizes their creativity and is fascinated by how it originates: "I believe that the twentieth/twenty-first century creative engineer can have more frequent and more sustained brainwaves than most of the rest—but there needs to be a scenery, a séance, a shoulder to look over."[2] This creativity has made the firm an asset to competition teams, and they are often brought in early in the design phase on projects ranging from houses and cultural buildings to high-rises and infrastructure.

Whether in their avant-garde ventures or in their more conventional projects, Bollinger and Grohmann practice a kind of Zen way of working—listening to and even "smelling" the architect's ideas, as Grohmann has expressed it, and maintaining a kind of fluidity with regard to withholding judgment about aesthetic issues. They lend support to the expression of the architect's design, devising structural systems suited to it but that are also beyond that which the architect could do on his or her own. They do not have preconceived solutions to a problem, set out step by step, as they begin. They approach each project with a fresh outlook, seeing it as a unique situation and set of challenges.

Both engineers are university professors and combine teaching at architecture schools with their practice, often meeting architects in that setting. Their work with Peter Cook, Wolf Prix, and Greg Lynn began in that way. Bollinger is a professor at the Universität für Angewandte Kunst (University of Applied Arts), in Vienna, where the firm also has a small office. As a lecturer for Peter Cook at Staatliche Hochschule für Bildende Kunste, Frankfurt-am-Main (City University for Fine Arts) in 1986, they formed a relationship that led to their association on the Kunsthaus in Graz, Austria, and also to Bollinger's connection with other contemporary architects. As a young engineer, he worked as assistant to professor Stefan Polónyi, a Hungarian engineer, at the University of Dortmund and wrote technical essays with him. Polónyi figured importantly in German Modernism and was one of the few engineers who participated in the Team 10 meetings in the 1960s. Bollinger was influenced by Polónyi's approach and inclusiveness that integrated engineering design, as well as the way he collaborated for years with architects such as O.M. Ungers. Bollinger's own focus is on the way that structure is about the making of space, or the forming of space through structures to enable the architect to achieve the desired effects. Architecture requires structure and the engineering has to support it. But when Bollinger looks at space, he knows the structure behind it, but what he sees is space; then he asks, "What does space mean for the structure?"

BELOW AND PAGE 52: Cinema Center, Coop Himmelb(l)au architects, Dresden, Germany, 1993–98.

OPPOSITE: BMW Bubble, Bernard Franken Architects, Frankfurt, Germany, 2000.

Grohmann, who was drawn to engineering because it dealt with built works rather than solely the empirical study of mathematics, started his career with the one hundred-year-old engineering firm Wayss & Freytag and has brought engineering into the digital age with his research on the issues of digital workflow, development of adaptable structures, and housing-construction systems. He teaches engineering to architecture students at the University of Kassel, focusing on bridging the gap between structural and architectural design at an early phase of the education process so that these architects in training will be better able and more apt to integrate structure into their designs. In addition to his research into digital workflow practices, which refers to the use of the same computerized data models from planning to production, Grohmann harnesses Finite Element (FE) computer software to calculate complex geometries and truss programs and assist architects and engineers in form finding for a design.

Computer programs have certainly had an impact on the way the firm works, allowing, for example, working design files, both two-dimensional and three-dimensional, to be shared. But Grohmann believes that the next programs can be developed beyond the one-sided entry systems now available, towards enhanced scripting abilities developed from the last generation of FE analysis as a design tool, and ways for computers to generate solutions for structural concepts. He notes that "We are trying to find rules behind the design; if there are rules you can program the rules and let

the computer run with them. We are doing it in our office and also at the University of Kassel. We have done a lot of scripting with Rhino and other programs and we established a bridge between Rhino and our FE program for special structures so we can define structure by scripting Rhino, check it with the FE program, and define fitness criteria. Then we can choose the fitness and can use a genetic algorithm."

Bollinger and Grohmann work from an office that is designed by architects Schneider + Schumacher, at Westhafenplatz in Frankfurt. The cantilevered bridge structure supports loftlike offices that feature triangular windows and an automated shading system. Schneider + Schumacher, who B+G have worked with since 1986, also designed the Infobox, erected in Berlin from 1995 to 2001, to display the history of the Potsdamer/Leipziger Platz construction site. The temporary 60-meter-long red-paneled steel box, which hosted over ten million visitors during the area's reconstruction process, was raised two stories above ground on pilotis, had 40-centimeter-thick, concrete-filled steel tubes, and was supported on bored-pile foundations. B+G used diagonal bracings to cantilever each end in a suspended and gridded system. A small, vertical Infobox was erected in 2002 in West Port, in Frankfurt, near their office.

The firm's partnership with Coop Himmelb(l)au, the Viennese firm of Wolf D. Prix and Helmut Swiczinsky, continues to transform both architecture and engineering culture. Their design for the Dresden UFA Cinema Center, the first contemporary building in Dresden after German unification (1993–98), won them the commission in a competition. Their scheme for the eight-cinema multiplex theater and its accompanying cafés explored ideas of defying gravity with jagged, "deconstructivist" spatial arrangements.[3] Angular structural steel girders allowed the glass facades of the central lobby to tilt at a 60-degree angle away from a singular volumetric core between two parallel concrete volumes spaced 18.5 meters apart, creating a hybrid structure. The glass facade, a crystal-like, seemingly weightless form, creates an illusion that it could defy gravity, however it is fully suspended from the concrete-and-steel structure wrapping around that core, in an irregular fashion. Another complex structural aspect of the design is the stairs that carry moviegoers to the four below-ground and four upper-level theaters. The metal stairs start in the interior atrium entrance and with an interlocking series of bridges wrap the central space, as though it were a maze within the structural form.[4] B+G's structural system was both efficient and economical: minimizing the ground-floor area and maximizing the cantilever used less steel. On the exterior wall screen the Palast projects advertisements for films shown inside. A café is suspended in a double-funnel-shaped cage within the atrium space—at once disconcerting and daring—a common goal of the engineers and architects. The engineers achieved the desired fluidity for the interior space by opening up the structure and minimizing the standard elements, depending instead on the exterior supports, similar to Coop Himmelb(l)au's 1960s-era projects for pneumatic shells in a solid interior structure. The Cinema Center is both fragmented and structurally expressive—its structure and formal qualities creating a dramatic effect: fragmenting views, disorienting vision, and skewing one's sense of well-being, while simultaneously embracing the space so as to engage the public.

Forms based on structural force fields is another area that B+G investigated with Coop Himmelb(l)au, with the unrealized idea of deforming the JVC Urban Entertainment Center roof in Guadalajara, Mexico (2011).[5] The form is based on the force fields of the steel lattice grid structure and derived from a digital modeling process using the computer program RSTAB. It followed imaginary gravitational forces to make an inverted force field landscape in a topographical roofscape.[6] To form a building based on force fields, the structure is organized not on principles of proportion, order, or symmetry but on the forces within, those that come from the structure itself. The stresses and strains of structure are similar to the investigations of natural structures completed by Robert Le Ricolais, who analyzed the holistic structure of bones to understand built works.[7] A similar force field–based design was employed in Coop Himmelb(l)au's Akron Art Museum (2006). The architects wanted the roof to fly—that is, to support the massive 40-meter cantilever on only one central column. To reduce snow load, they made the ends permeable.

Not all of B+G's projects are feats of engineering; the engineers are also masterful at slightly calibrating structures to solve programmatic issues. One such project is the unbuilt extension of the 2002 Hamburg Fair pavilion, designed in conjunction with Albert Speer & Partner with Schweger and Partner. In the extension, the roofing system is also a design element. Rather than build separate pavilions, they had proposed to enclose the main area under one roof. In a modular, 20-by-20-meter slab raised 4 meters at the edges, four slabs in a variegated pitch pivot and tilt around a central point and are repeated in a 40-by-40-meter grid. The separation between the square steel structures allows light to enter the space through triangulated slats, achieving the desired flexible open shed but in a nontraditional, infinitely repeatable space-truss design.

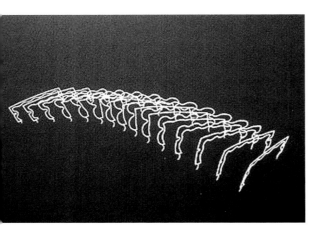

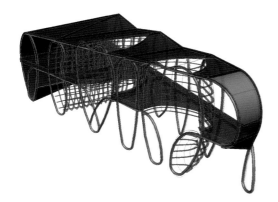

B+G have also enhanced and enabled the work of German architect Bernhard Franken, whose design for three smaller-scale topological branding structures for BMW—the Bubble (2000); the Dynaform with ABB Architects (2001), in Frankfurt; and the sculpture Take-Off in Munich (2003)—merge conceptual ideas of self-supporting forms with those of new computer-assisted design and fabrication technologies that make transparent and curvilinear shapes possible.

The shape of the Bubble (approximately 24 by 16 by 8 meters) was based on drops of water and their continuous, uninterrupted smooth surface, an irregular volume, neither a sphere nor an ovoid. The design's translation from virtual model to fabrication of elements to physical building was effected through a digital workflow using parametric modeling.[8] Jointly, the engineers and architects decided on methods of fabrication and construction, but as construction proceeded, B+G determined that additional support was needed. The structural aluminum ribs that support the slump molded, heat-formed acrylic skin, were recalibrated and refabricated so that the skin and the ribs could work in concert, with the ribs taking the primary loads and the shell acting as a stabilizer. These changes led to the development of the concepts for the Kunsthaus Graz structure, showing the design influence of the engineers as they form their own approach with different architects.

In the BMW Dynaform pavilion, the load-bearing structure became a self-supporting form, without extra beams and ties. Franken designed the pavilion's shape to simulate driving inside the pavilion, devising a Doppler effect to organize the interior surface. The form is based on force fields of the site context, as well as those of the building's physical forces, simulated on the computer. This became the architect's "master geometry" that guided the project. The engineers realized that the skin could not be self-supporting. They incorporated a series of fifteen dyna-frames (thus the name Dynaform) in the axes of the structural section so the structure could grow into a long rectangular tube-shape that performed as hollow-box steel girders. Lattice structures with additional short tubes connected to the frames into tangent arcs in polygonal tessellation with all the frames welded by hand. A membrane textile in tension against parallel lines, and around the framework, tested with a FE analysis program and in the 3-D modeling process, created the desired translucency.

B+G is working with Japanese architects SANAA and engineer Mutsuro Sasaki on a six-story office building for the Novartis campus in Basel, Switzerland (2007), that faces the main street and includes a large green space at the entrance. B+G devised super thin reinforced-concrete structural slabs for the 84-by-22.5-by-22-meter volume supported by structural walls spaced 10.5 meters on center. This achieves the desired column-free and unobstructed visual transparency through the building, making it appear as a series of floating louvers. Windows are finished with a coating to reflect the greenery and the sky, so that nature is absorbed into the building's skin and the sunlight is controlled. Both offices and support spaces are located along the periphery to receive natural light, opening up the center space for a large interior courtyard and challenging the engineers to a proposition for the least amount of materials and maximum openness.

The Slavin House—designed by Greg Lynn for a Venice Beach, California, site, but as yet unbuilt—folds inside and outside rooms into a singular porous environment that occupies the entire triangular site. B+G's challenge here was how to express the spatial volume within structure. They devised a one-story-high occupiable structural truss of two continuous extruded steel

tubes that are bent and looped to function as horizontal and vertical members, beams and pilotis, structure and decorative elements. The integrated structure allows for a one-hundred-foot-long ground-floor living area to be partially enclosed and yet merge with outdoor spaces. Light courts perforate the upper level. Each element of the house serves more than one purpose: material and surface continuities make volumes both voids and solids inside and outside, and the curvilinear basket structure both supports and creates hollow courts. This flowing continuity of upper and lower levels, of roof and ground, engenders a new kind of porous and fluid house that folds together space, frame, skin, and envelope.[9] "The whole 'blob' conversation as well as the 'undulating' conversation and the 'exotic surface' conversation—they will all surely pass through the B+G office, but we should realize that the office knows more about the spirit and the letter of such ideas than many of their adherents," says Peter Cook.[10]

Grohmann notes that he expects "new structural systems and new structures that go far beyond all that we have learned and seen until now." Since the days of Carl Culmann (1821–1881, German engineer, author of *Die Graphische Statik*, 1865, and professor of graphic statics at the Polytechnical School of University of Zurich), only structures that engineers could analyze have been built. "New software tools will allow us to mix different types of load-transfer systems that combine elements that are already known in a new and hybrid way." As Peter Cook has said, "Whilst rarely pronouncing that an idea is impossible (or stupid, or inappropriate), B+G gently coerce you away from the inept towards the appropriate. Then to something more: towards the desirably appropriate. Then to something more still: the unexpectedly and desirably appropriate."[11]

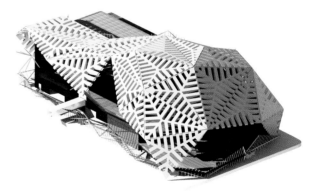

ABOVE: Novartis, SANAA, Basel, Switzerland, 2005–7. Construction "bubble deck" (left); model (top right); Théâtre Mariinsky II, Dominique Perrault, St. Petersburg, Russia, 2006–8 (bottom).

OPPOSITE: BMW, Dynaform, Bernard Franken Architects, 2001. Diagram (top); installed in Frankfurt, Germany, 2001 (middle); Slavin House, Greg Lynn FORM, 2007 (bottom).

BMW WORLD
MUNICH, GERMANY

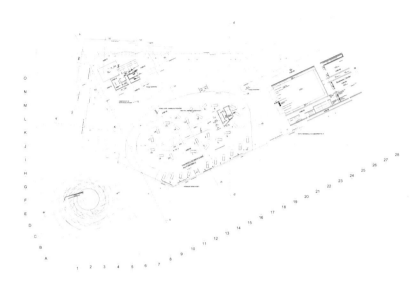

ABOVE: Plan of car display floor.
BELOW: Section showing the programmatic insertions.
OPPOSITE: Structural wireframe diagram of the Double Cone.

When, in the 1960s, the architects Coop Himmelb(l)au created living units with pulsating special supporting structures such as the pneumatic Villa Rosa (1968) and the mobile, inflatable "Living Cloud" (1968–72), they believed that in order to expand the field and be ephemeral "architecture must remain in a state of unrest . . . Where architecture had fantasy" it could be something else as "buoyant and variable as clouds."[12] But these early experiments often lived on only in small-scale models or as drawings, and it was only in the 1990s, with the collaboration of engineers such as Bollinger and Grohmann, that some of these concepts have been built. And as static structures they are as vibrant as in the original conception.

One such project is the BMW World in Munich, a car distribution and entertainment center adjacent to the 1972 Olympic Stadium, the 1972 BMW office tower, and the 1973 BMW Museum. Commissioned through a competition in 2001, Coop Himmelb(l)au designed the new complex focusing on the idea of creating a large, open, multifunctional space, in a new shape of space, horizontally oriented under a fluid roof structure, in contrast to the existing vertical tower.

The undulating roof connects the separate building functions—the Double Cone, an exhibition and event space; Premiere, with its lounges, cafés, and shops; Forum, a 600-seat theater, and the main Hall, a lobby and shopping mall—in a 23,000-square-meter space. An immense three-story underground space of 45,000 square meters is the car-storage garage, where new owners pick

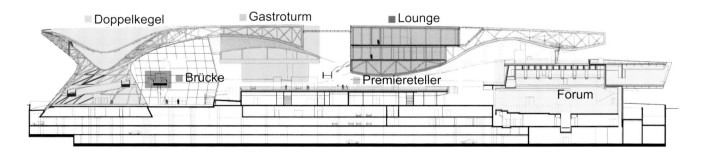

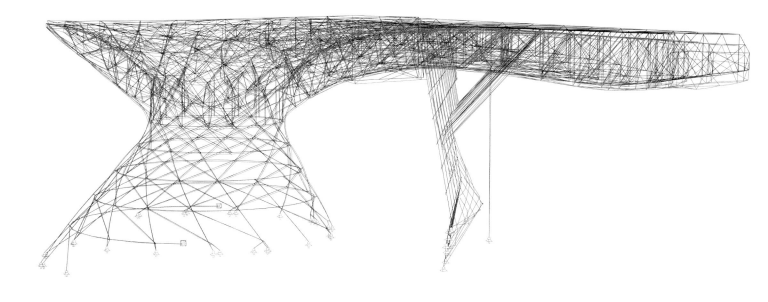

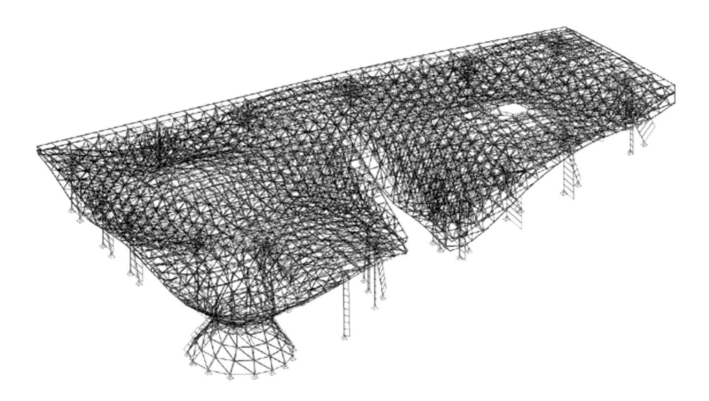

up their cars. To illustrate the conceptual expression of the ephemeral in solid form, the engineers designed a cloudlike roof 200-by-120 meters in steel and glass that is supported at one end by a space that is an unusual double cone 28 meters high by 45 meters long, with additional supports from eight sculptural concrete columns and five cores. The deformations in the roof—the dips and curves—are calculated using the concept of imaginary gravitational forces to achieve structural balance. At specific points where the structure opens up and voids are contained, the public spaces and offices are housed. The steel girder space-frame structure of the 25-meter-high roof has a pocket of space that contains solar cells to make a passive-solar section of the building. The network structural mesh is one that provides as much uninterrupted free space as possible. Above, a cushion of space was created through simulation reaction of the underlying area. The engineers worked to shift the structural system with each rendition of the architect's design, indicative of the digital flow of work between the collaborative team. In the 13,500-square-meter glazed facade, maximum transparency was achieved because the facade is suspended from the roof structure. A modified post-and-beam structure is folded to accept vertical roof movements without any structural joints, resulting in the fluid, cloudlike formation imagined by the architects. As Peter Cook said, there is a "sheer symbiosis of the BMW World building: truly experimental and totally original. The product of sheer virtuosity: both inventors are the single virtuoso."[13]

ABOVE: Aerial view of construction site; structural framework of cone; installation of glazing (top row); model.

OPPOSITE: Structural diagram of the wind, snow, and live loads (above); models of the main glass facade and atrium (below).

KUNSTHAUS GRAZ
GRAZ, AUSTRIA

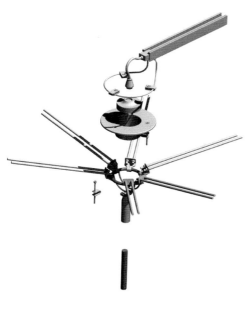

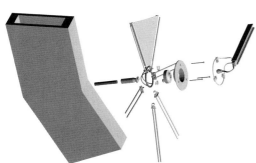

ABOVE: Steel connections for the Plexiglas facade system.

RIGHT: Axonometric drawings of floor plans.

OPPOSITE: Northwest facade at night, showing the BIX florescent lighting system.

In 2000 a jury for an international competition selected Peter Cook (formerly of Archigram) and Colin Fournier's design for the Kunsthaus Graz, to be situated on an irregularly shaped site in the historic center of the city. It was completed in 2003. The firm, known as Spacelab Cook-Fournier Architects, sought a continuous surface that would blur roof and wall resulted in a biomorphic object, a sinuous building set off from, but that still related to the adjacent historic cast-iron buildings.

The bulbous volume envelops two levels of gallery spaces in a stacked structure, with the steel grid of the ground floor supporting the loads of the bubble so that it appears to float. The engineers' challenge was to maintain an open structure by keeping the space flexible for what they saw as potential future uses. They employed only five columns, with two structural "tables" stacked on each other. The lower table was a steel grid and the upper one in the bubble was an exhibition area. As Peter Cook described, "The columns progressively became fewer in number and slimmer in girth. Similarly the skin became more and more voluptuous as it became more convincing."[14] Other structural supports include two bean-shaped concrete cores integrating structure and mechanical systems in access channels.

The museum includes new-media spaces, multipurpose areas, a library, offices, and an underground garage for the surrounding commercial area, resulting in a six-story building topped with the "Needle," a 40-meter-long horizontal shaft beam of steel and glass that serves as an observation point. Interior circulation between levels is provided by two upwardly moving 30-meter-long travelators, or moving sidewalks, which create a strong diagonal across the open space, and a stair down. The uppermost access leads to the Needle and brings visitors in and through the space to the outside via other stairs, balconies, and passages of this porous building.

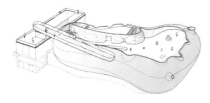

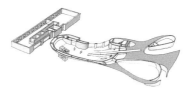

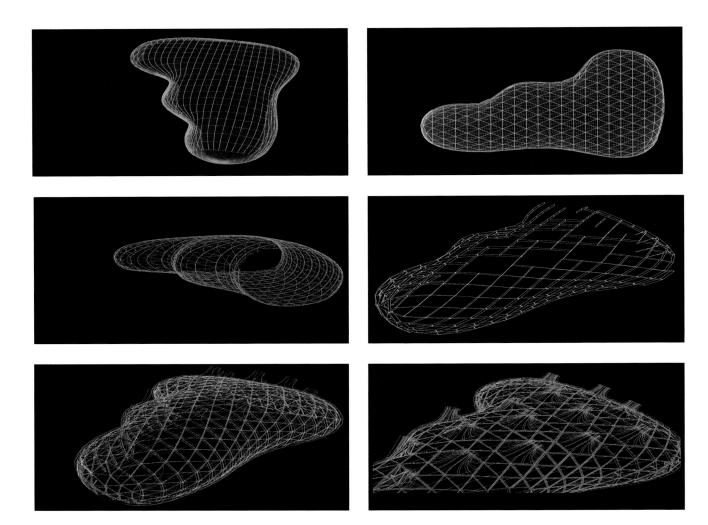

B+G worked with the architects from the outset, corresponding through digital tools to develop the structural optimization. The team created a three-dimensional model using Rhino software, which they could then manipulate when designing the structure with parametric modeling. The intensity and focal point of the structure is the double-curved bubble-shaped skin of the main volume. There are 1,300 individually and thermally formed Plexiglas panels whose form is parametrically modeled to follow the desired shape of the building, as in Bernard Franken's BMW exhibition stand (Frankfurt). The greenish blue Plexiglas casts a second skin over a steel roof of welded polygons in parallel arched rectangular steel box girders. A plastic weatherproofing and fireproof membrane, with a separation between the two, forms an interior surface. The sleeves are welded where the support struts penetrate the volume, and the panels are held in place with stainless-steel bolts at each corner. The bubble

ABOVE: Gallery.

OPPOSITE: Development of the structural steel diagram (above); connection detail of the facade system (below).

receives light through funnel-like openings, or "nozzles" on the roof, similar to north-facing factory light wells, providing a small amount of daylight within the upper spaces. The 60-by-40-meter Plexiglas roof/wall, at first envisioned to be transparent, was then transformed into a media display, called BIX. The firm realities:united inserted circular fluorescent tubes between the facade's two layers, and those tubes are animated by computer, creating a synergy between media and architecture.[15]

As Colin Fournier said, "The smoothness of the building's resultant double curved surfaces has gentle and cuddly connotations which, combined with the peculiar nature of its nozzles, its multiple snouts and eyes, have led to the nickname 'Friendly Alien.' Indeed, the building seeks, through its appearance and modus operandi, to be a friendly institution that is easily accessible to the public and adopted by the people of Graz as a strange but familiar part of the normal life of the city."[16]

BURO HAPPOLD

Sir Edmund Happold (1930–1996), or Ted as he was called, founder in 1976 of the firm Buro Happold, believed that "a world which sees art and engineering as divided is not seeing the world as a whole . . . It is technology that is creative because it gives new opportunities. Historic ideas of art and culture can entrap. It is technology that frees the scene."[1]

Happold embraced both creativity and technology in a way that led to new, overarching concepts in design engineering.

Happold intertwined aesthetics and technology with "an approach to design whose roots are not dependent on visual precedents. I am referring to engineering design as a technological idea, with its own aesthetic."[2] He related ideas of the aesthetic of structures to that of mathematics and form and the beauty that can be found in mathematical solutions. For him, "structural design is primarily concerned with the choice of form; the forces on that form and the analysis of its behavior follow on from that choice. The whole process is influenced by the need for feasibility of execution, as success is proved by practicality." Happold thought that structural design had to "achieve elegance as well as value; elegance in the mathematical sense, meaning economy as well as appropriateness. Appropriateness (or function) + economy = value."[3] And he turned this into a diagram that he often drew to express this quality.

Cogently, Happold believed that creative engineering stimulates good architecture. He "pursued efficiency and economy while conscious of the artistic, social, symbolic, and intellectual issues of our built environment. He understood the need to reduce architecture and science to the practicality required of construction, where meticulous engineering and a sense of scale and ingenuity of detail add to the nature of the architectural solution."[4]

Happold also encouraged and demanded from his staff collective decision making for engineering and design in a group process. And he did not exclude himself from this process. He was in "an exalted position but was always approachable, welcoming, and engaging."[5] Much of this could be attributed to his Quaker background and his holistic worldview, which extended to his practice and to structural engineering in general, which he saw as a social responsibility. This quality he imbued in his firm, Buro Happold, which carries on his name.

After studying engineering from 1954 to 1957 in the Department of Civil Engineering at Leeds University, he traveled to Finland where he met Alvar Aalto and was inspired by his design aesthetic. Happold began working at Ove Arup & Partners in 1957 with the engineer Povl Ahm, who was the lead designer of the screen frames for St. Michael's Cathedral, built adjacent to the ruins of Coventry Cathedral (Sir Basil Spence, 1955–60). Happold studied architecture in the evenings. Further inspiration came from Fred Severud's engineering for Eero Saarinen's projects such as the Ingalls Ice Rink (New Haven, 1958) and Dulles Airport (Washington, D.C., 1958–62), and Happold went to work for him in New York in 1959. Severud's unexpected building solutions achieved by means of straightforward engineering principles—such as an out-of-plane stiffness to a grid of steel cables for the saddle-shaped suspension roof for the Raleigh Arena (Matthew Nowicki, 1952)—opened up Happold's mind to nonstandard solutions. Happold returned to Ove Arup & Partners in 1961, continuing to work on the buildings of Sir Basil Spence.

Many engineers aspire to design lightweight structures because they offer a chance to explore form, efficiency, and material. This was especially true for Ted Happold, who became enthralled with German architect Frei Otto's long-span designs for temporary exhibitions.[6] In 1964 he visited Otto's Lausanne Swiss National Exhibition pavilion (designed with Rolf Gutbrod)—a tensile roof with a peaked shape and saddle surfaces with a rope-net tension structure enclosed in plastic with lattice steel masts—and the 1966 Montreal Expo buildings (with engineers Leonhardt, Andrä and Partners)—a free-form single membrane of irregular plan and heights that rise and fall to the ground with translucent fabric. Otto was known for his examination of the link between form and structure and of the potential of materials to find their own shape; his discoveries and analyses were detailed in his 1954 dissertation, "The Hanging Roof." He had completed smaller tensile structures, but larger ones were more difficult because the fabrics available at the time were not strong enough and the calculations were complex. Otto explored the use of cable-net systems, which could separate the load-bearing and envelope functions of a roof using a mesh of steel cables to support a woven polyester fabric that did not carry any loads. This work was based on his study of soap film and its ability to achieve its form with the least amount of surface area.[7]

By 1967 Happold had become head of Structures 3, one of the four design groups at Ove Arup & Partners. Among his colleagues in the group was Ian Liddell, who had joined Arup and worked on the roof of the Sydney Opera House from 1960 to 1962; completed postgraduate courses at Imperial College to study prestressed concrete and shell structures; and then returned to Arup, working on the Hotel and Conference Center in Riyadh, Saudi Arabia (Trevor Dannatt & Partners, 1966-73). Liddell, who later became a partner of Happold, completed the geometry of the complex fan-shaped roof, the structure of the foyer, and the column support to the space-frame roof of the conference center—demonstrating how structural design, here oriented to thermal movement, can be articulated to enhance the architecture. To complete the intensive calculations, the group used a huge mainframe computer at Manchester University.

In 1971 Happold's group proposed that Arup enter the Pompidou Center competition, in Paris, with architects Richard and Su Rogers and Renzo Piano. With Happold, the architects developed the idea of a gigantic interactive information machine. After winning the project, the team opened a project office in Paris, with Happold at the lead. His main contributions included those submitted for the competition, such as exposed-steel scaffolding, steel castings, water-filled interior tubes, and long spans for the steel structure. While his involvement continued through the period of contractual negotiations with the government because Happold had established a relationship with Robert Bordaz, the client,[8] he was called back to Arup's London office, and Peter Rice and Geoffrey Wood, who were also part of Structures 3, carried out the scheme, leaving Happold's contribution to the project largely unrecognized.[9]

Happold's continued interest in lightweight structures prompted him to start a Lightweight Structures Laboratory in 1973 with Ian Liddell, Vera Straka (who now practices in Toronto and teaches at Ryerson University), Peter Rice, and Michael Dickson (who became the chair at Buro Happold after Happold died in 1996).

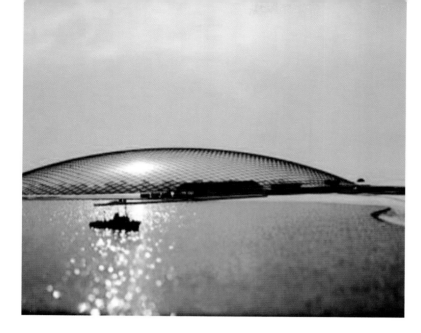

This research group corresponded to Otto's Light Structures Institute at the University of Stuttgart.[10] Tent structures, compression supports, and tension-loaded membranes were researched with small teams and with students. The Lightweight Structures Laboratory worked on the design for the master plan of Riyadh University, with architect Karl Schwanzer; a 20-acre stadium for the University of Jeddah in 1976, with Arup and Happold; and the Grand Mosque, which won an Aga Khan award in 1974. One fascinating futuristic concept—similar to Buckminster Fuller's Old Man River, an umbrellaed-town concept for St. Louis (1973), or his Dome for New York City (1960)—was for 58 Degrees North New Town, a bubblelike structural envelope designed in 1971 with Kenzo Tange and Frei Otto with Arup & Partners to cover an arctic city in order to maintain a temperate climate. While at Arup, Happold was the principal engineer, in 1975, for the timber gridshell of the Mannheim Bundesgartenschau Multihalle (National Garden Festival Hall). He introduced the use of Ethylenetetrafluoroethylene (ETFE), a plastic foil, as a cladding material, and developed software for form finding and analysis of fabric structures that has led to recognition of his firm's later work in long-span fabric structures, including London's Millennium Dome.

In 1976, when Happold was appointed professor of building engineering at Bath University, he left Arup to start his own firm, Buro Happold, with seven other engineers from Arup, including Peter Buckthorp, Terry Ealey, Ian Liddell, Rod Macdonald, John Morrison, John Reid and Michael Dickson. He aspired to coordinate an integrated education curriculum at the University of Bath so that designers and engineers shared knowledge. He continued his independent research by organizing the Centre for Window and Cladding Technology and the Wolfson Air Supported Research Group, which completed studies with architect Arni Fullerton for an air-supported structure with ETFE foil cushions that would form a roof covering a 35-acre site in northern Alberta, Canada, for a village of 2,000 people, but the idea was never implemented.[11] The research into fabric, tensile, and air-supported lightweight structures was ongoing. Explorations of materials and structures were done in collaboration with fabric companies such as DuPont, whose development of Teflon and PVC coatings and PTFE (polytetrafluoroethylene), a vinyl polymer coated glass, were in turn enhanced by the engineers' structural research. Ian Liddell used ETFE foil cushions rather than a glass roof over internal streets for the Chelsea and Westminster Hospital (Sheppard Robson, 1993). The foils are placed in frames and clipped on to a substructure, which contains the gutters.

Buro Happold continued to collaborate with Arup through the 1970s on projects such as the Mannheim Hall, also in conjunction with Otto. Mannheim was a timber lattice shell structure composed of 50-by-50-millimeter laths of western hemlock of unprecedented spans and irregular,

ABOVE: 58 Degrees North New Town, concept for an arctic city in Canada, 1971 (top); ETFE cushion roof, Chelsea and Westminster Hospital, Sheppard Robson, London, 1993 (bottom).

OPPOSITE: National Garden Festival Hall, with Arup and Frei Otto, Mannheim, Germany, 1975.

PAGE 66: Japan Pavilion, Expo 2000, with Shigeru Ban and Frei Otto, Hannover, Germany, 2000.

double-curved forms. The success of the entire timber lattice depended on the elasticity of the spring washers at the node joints between each pair of laths, in an illustration of Hooke's Law.[12] The lattice shell was unstable at first because of shear issues, and the dome would have collapsed, so the team reengineered it, increasing stiffness and buildablity to get the right form and connection details around the boundaries: "true engineering," explains Liddell, who worked on the project. This trajectory of the structurally driven form is continued in the firm's more recent work, such as the Weald and Downland Open Air Museum.

With architect Theo Crosby of Pentagram, Buro Happold, in one of the first prominent jobs for the new office, designed a tent for the 1976 British Genius Exhibition in Battersea Park. Ian Liddell designed a large column-free area, drawing on the concept of air-supported structures developed by the company Birdair, who Buro Happold often worked with. The firm also had designed numerous projects in Saudi Arabia, and there the tent structures, an intensified version of the Bedouin tent, became a natural solution. For projects such as the Diplomatic Club for the Saudi Arabian Foreign Minster in Riyadh (1985) with Frei Otto and Omrania & Associates of Riyadh, they created curvilinear concrete and limestone walls that enclosed a banquet hall, entrances, and gardens, which led to outer bubbles reserved for sports, clubs, lounges, and accommodations in the shape of a rose. The Doha University, in Qatar (Kamil Eli Khafrawi and Renton Howard Wood, 1975–85) incorporated environmentally appropriate wind towers inspired by the vernacular architecture, meeting both the structural and environmental sustainability needs for this desert complex. Continuing to explore structural forms inspired by nature, Buro Happold developed branch structures to support shading devices in the desert, imagining entire regions under shade. This innovation was then transferred to temporary structural uses such as Frei Otto's cotton umbrellas for the rock band Pink Floyd's 1977 American tour. For the Quba Mosque, in Medina, Saudi Arabia (1984–86), they created a field of folding parasols of a fine silk-like Teflon PTFE, Tenara, which was pliable as well as fire- and ultraviolet-ray resistant.

Ted Happold believed that, in "contrast to the common perception of engineers as technicians, the results of their work are as distinctive and diverse as the work of different architects."[13] However, he did maintain a distinction between the two professions, and in a talk for the Institution of Structural Engineers in 1986, "Can you hear me at the Back?" he alluded to the fact that in reality structural engineers take the subordinate role in designing. "We must explain that we are not 'just the structural engineer' but jointly designers, bringing together a knowledge of structure, materials and construction to the problem."[14]

In 1982 Buro Happold began to work in the United States with Nicholas Goldsmith and Todd Dalland of Future Tents Ltd. (FTL) on temporary and recreation projects using cables, masts, and tensile membranes, including the Boston Harbor Light Pavilion (1997) and numerous tent coverings for stages. After ten years the two firms combined their New York practices, becoming FTL Happold employing both architects and engineers on staff. In 1998 the firm split again, and Buro Happold formed its own office in New York.

Buro Happold also developed Tensyl, flexible write patterning and mapping software specifically for the design of tensile structures. The fabrication geometry module allows for membrane-cutting patterns to be generated (like fabric patterns), so that the related geometry for components for masts and plate angles can then be adjusted on-screen. The visualization models can be viewed from numerous perspectives. Analysis for materials, lighting, and textures can also be added. First used for cable nets and fabric structures, and then expanded for surface structures and amorphous forms, Tensyl can map flat panels on a curved surface, in both nongeometric and Cartesian forms. This mapping ability placed the firm in demand with curvilinear-oriented architects such as Greg Lynn, who designed the Korean Presbyterian Church, in Queens, New York (1995), by using computer programs to produce interactive models in many iterations. The New York office also teamed up with designer Chuck Hoberman on the

ABOVE: Diplomatic Club for Saudi Arabian Foreign Minister, with Frei Otto and Amrania & Associates, Riyadh, 1985 (top); colored glass tent with paintings by Bettina Otto, 1986 (bottom).

OPPOSITE: Millennium Dome, Richard Rogers Partnership, London, 2000. Diagram showing the steel masts and cables (left) and installation of the PTFE coated glass fiber roof (right).

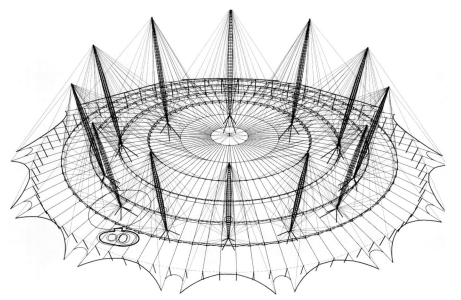

Hoberman Arch, an aluminium paneled, unfolding arch that spiraled out and up to unfold for the stage of the 2002 Salt Lake City Winter Olympics.

The experimental nature of Buro Happold's work is at the height of expression in tensile structures such as the Millennium Dome in London (Richard Rogers Partnership, 2000), the design of which evolved from British marquis tents. Happold partners Ian Liddell and Paul Westbury of the Bath office had completed a tent study for a standardized system and discovered that flat panels held in tension, like circus tents, were the simplest forms. They designed a trial project for a 93,000-square-meter tent for the religious group RSSB (Radha Soami Satsang Beas), and they then applied the chief characteristics of its design—a unified, flat surface and tension cables—to the Millennium Dome, so that in the end it had a low level of architectural input. In fact, it received an engineering award rather than an architectural one. The dome has a PTFE coated glass fiber roof, tensioned with steel cables radially distributed to the edges, forming a covered 100,000 square meters of space. It is 320 meters high with 12 steel masts. Says Liddell, "The concept for the cable structure was rushed because we had six months from the start of the design to send drawings to tender, and in the end all the drawings were the engineers' not the architects'. Mike Davies of Rogers was actually happy about it, and said, 'Keep going.' " It is a breakthrough project for fabric. As part of revitalization plans for the Greenwich Peninsula, the Millennium Dome will be adapted as the 02 Arena for a second life as a 20,000-seat sports and entertainment center.

Combining its work in shell and tensile structures, Buro Happold has designed a new, underground Stuttgart Train Station (Ingenhoven Overdiek and Partner, 1997 competition; anticipated completion date, 2013) as part of the Deutsche Bahn's realignment of the rail lines so the station can be part of the intra-European network. Above ground, Buro Happold, again with Frei Otto, has engineered a roof deck that will connect to a new linear public park. An organically shaped, steel reinforced, poured-concrete shell with a calyx section creates columns that terminate as skylights, or "eyes" that let in both light and air, eliminating the need for mechanical systems. The apertures' ring beams will transfer the loads to the columns. Earlier innovations such as gridshells, or single-layer articulated reticular shells, were applied to experimental materials such as the cardboard tube Japanese Pavilion for the Hannover Expo (Shigeru Ban and Frei Otto, 2000). The pavilion, made of 22-millimeter-thick paper tubes lashed together with plastic straps, was assembled flat and lifted up. Liddell notes that "the cardboard structures are a bit decorative, since the paper board is not structurally capable of doing the job;" in the end the checking engineer on the Hannover Pavilion asked for more steel. In Ban's Pier 54 Nomad Pavilion (New York, 2004), where cardboard was used to form columnar and truss structures, there were difficulties with glue stability because cardboard creeps over time. Other cardboard projects include the Westborough School in Westcliff-on-Sea, Essex, England (Cottrell and Vermeulen Architecture, 2004), and a Pavilion for the Museum of Modern Art in New York (Shigeru Ban, 2000).

Craig Schwitter, a partner in the New York office, explains, "The structural work today is not a prescriptive engineering that has been done before and that is known, it is a performance based design, which involves a great deal of testing and it is a bit risky." Performative engineering—developing a structure from an understanding of its material qualities and how it works, moves, and behaves, rather than from expected external forces—often relies on computer simulation and is beyond what the building code can imagine.

WEALD AND DOWNLAND GRIDSHELL OPEN AIR MUSEUM
SINGLETON, CHICHESTER, ENGLAND

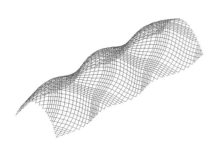

The Weald and Downland Gridshell Open Air Museum, designed by Edward Cullinan Architects with Buro Happold engineers in 2002, follows from the structural exploration the firm initiated with Arup for the Mannheim Garden Festival Hall and the Hannover Expo Pavilion. It was the first double-layer timber gridshell in the UK.[15]

The museum consists of a collection of timber-frame vernacular buildings that have been assembled on the property and is also a conservation center for the study of timber-frame construction. The workmanship fits the program both pragmatically and didactically as the next generation of wood technology. Similar in concept to an airplane hangar as a holistic space, but here a doubly curved shell in a triple wave form, the gridshell structure is composed of four layers of 35-by-50 millimeter green oak laths on 1 meter centers. The wavy, ribbon-shaped structure is 48 meters long and between 16 and 11 meters wide. The wood is strong but supple, with a high moisture content, which gives it flexibility, allowing the frame to be shaped at the time it is erected. The oak was selected based on its availability from sustainable-harvest companies in Normandy, and had a higher bending strength than other available wood. Defective sections of wood were cut out and the pieces reconnected with finger joints to form 6-meter lengths, with scarf joints to make 37-meter-long laths for the lattice and 50-meter-long longitudinal rib laths, so that there was very little waste.

The process depended on collaboration and teamwork. Among the participants was the Green Oak Carpentry Company, which created the connecting device for the laths that allowed for movement and stability. The system, as an industrial invention, was patented by all the teams on the project—architects, engineers, carpenters, and the museum. It consists of three plates with a pin in the center that are bolted down at the corners of the plates once the shell form is in place to make a shear stiffness. The outer two laths can slide in three dimensions, and the middle pairs only in two, to maintain the grid shape. Unlike the Mannheim shell, which was constructed on the ground and raised, this one was constructed flat on a scaffolding support seven meters above ground and then lowered down. The supports were only removed when the shear blocks had been screwed into place between the layers of the shell

ABOVE: Computer simulation of forming the timber gridshell.

RIGHT: Sketch of section and plan of the lath layering. system.

OPPOSITE: Forming the gridshell system (above); south facade (below).

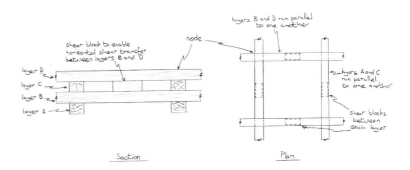

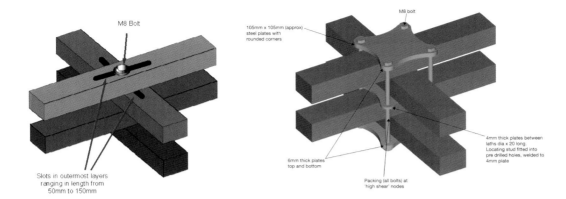

ABOVE: Diagram of the slotted hole option (left) and the nodal connection (right).

BELOW: Gridshell uses traditional and locally available materials.

OPPOSITE: View from west into the main space.

so that the parallel lines of laths acted compositely following forces from the structural analysis, to lower toward an arched cross section and were then bolted to the edge beam. Cantenary arches were used for the ends of the shell as awnings.

Vertical planks of western red cedar clad the lower walls, leading to a large clerestory and to the single-sheeted roof, which consists of four layers of steel mesh stapled to the plydeck. The concrete cellar level is sunken into the chalk hillside and houses the artifact archive relating to wood production, such as tools and historic building elements. The sloped site lent itself to the architect's juxtaposition of structure and superstructure, of solid wall and a transparent lightweight, polycarbonate clerestory/skylight. A floor structure of laminated beams on a central row of Douglas fir paired, Glu-lamimated columns, support a timber-plank deck between the lower and upper level of the arched space. It cantilevers over the reinforced-concrete walls of the archives area to support the gridshell; educational workshops are held inside the space. The building is ecologically sound and innovative in its use of materials and in its integration of structure and form.

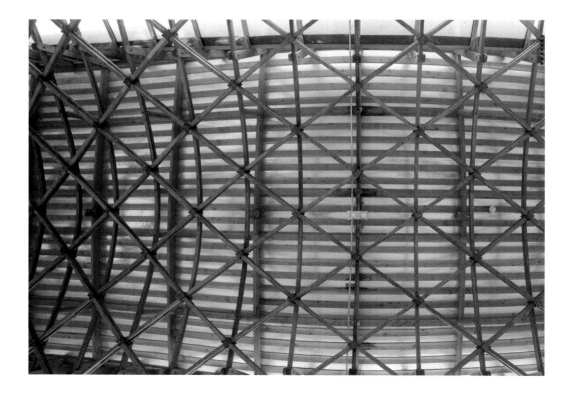

THE BRITISH MUSEUM GREAT COURT
LONDON

Buro Happold's reputation for being able to cover great expanses of enclosed space—as demonstrated by the Millennium Dome (2000)—brought the firm an opportunity to work with Foster + Partners on the restoration and adaptation of a long-obscured grand courtyard at the center of the British Library in London. Foster's firm won the commission in a 1994 open competition. The program required a new roof covering for what would be a 6,700-square-meter public gathering place filled with galleries, shops, and cafés. Although the space was to be formed primarily out of Robert Smirke's 1823 neoclassical courtyard and Sydney Smirke's 1857 Reading Room, which was also being restored, Foster increased the space by 40 percent by removing various incremental and less significant buildings and book storage. The Reading Room, which never had a finished exterior because book stacks surrounded it, was newly clad in Galicia Capri limestone; this centerpiece of the courtyard is reached by two elliptical baroque staircases. Now called the Queen Elizabeth II Great Court, it was the largest covered courtyard in Europe at the time of its completion in 2001 and continues a trajectory from Foster's design of the Reichstag Dome in Berlin.

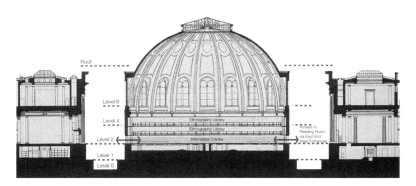

TOP: Elevation with historic Reading Room.
ABOVE AND OPPOSITE: Glass and steel roof structure.

The project required a balance between the architect's desire for a lightweight roof and the need to respect the historic structures, including maintaining the sight lines to the dome and the off-center Reading Room. Mike Cook and Stephen Brown of Buro Happold designed the roof in glass and steel as a lattice shell structure, and without visible supports so that it does not intrude upon the historic courtyard facades but keeps the space and historic structures independent. Form-finding techniques with soap-film analysis were used to define the roof, the shape of the individual panes, and the overall structure. Initially, a larger inflated bubble shape was developed by Chris Williams of the University of Bath, who consulted with the engineers on the stress levels, adjusting it to fit. The supports were fastened at the exterior edges of the roof structure, as well as to the interior drum of the historic Reading Room. A ring of twenty slender concrete-filled steel columns is installed around the Reading Room. Tucked between the original wall and the new stone cladding, the columns support the center of the roof. The 457-millimeter-diameter tubular steel columns meet the required fire rating and reach the 19 meters from the floor to the gallery. To support the dead load of the glass roof and the live load of water or snow, the engineers designed a new perimeter ring beam, on sliding bearings, independent from the courtyard walls.[16] One could consider the roof as a free-flowing surface of metal and glass mesh that then warps to have uniform strength and stiffness and resistance to wind and snow loads. In the snow gallery, between the new and existing domes, a new concrete slab acts as a stiffening diaphragm that prevents the roof from spreading laterally, balancing the thrusts from the opposite sides and avoiding applying lateral load to the existing quadrangle buildings.[17] The roof is formed by radial hollow roof members spanning between the Reading Room and the quadrangle buildings, tapering from 180 millimeters at the edge to 80 millimeters at the center and spanning up to 40 meters. The roof curves to a radius of about 50 meters, like a dome. Two opposing spirals of cables interconnect

the steel elements, so the system works as a shell. The specific curvature allows the structure to be based on arch compressions.[18] A series of triangular units of steel and glass generate out from the circular center to the rectilinear perimeter, radiating out in two interconnected but opposing spirals in toroidal geometry. Steel latticework supports 3,312 triangular glass panes that bend with the geometry dipping down in the center, around the edges of the circular Reading Room. The largest steel members are the corner external trusses that work in bending and compression, providing the roof with additional lateral stabilization, as does the cross bracing installed behind the portico facades, helping to withstand thermal change as well as movement from wind loads. Reinforced concrete parapet beams support short steel columns at every other of the 1,800 nodal points that connect to the beam system and hold the glass in place. The steel fabrication company, Waagner-Biro—which also worked on the Reichstag Dome and the Sony Center Forum roof, both in Berlin—developed a system to assemble the triangular steel pieces as a series of prefabricated ladder beams from a scaffolding, or deck, that covered the court. In a computer numerical control (CNC) fabrication process, each steel frame and glass panel was separately produced.

Slight differences emerged over time from the competition entry to the final design scheme. Originally, the roof material was to be PTFE fabric, but when glass became the material of choice, more structure was needed. The roof evolved from an orthogonal grid to a three-way lattice, and the geometries were developed from the Happold Tensyl computer program, to integrate asymmetry into the Reading Room's wrapper. Other issues included excavating six meters below the basement while the museum remained open, and optimizing solar gain from the double-glazed and fritted glass roof panels, and effecting natural ventilation. The result is a flowing vortex that appears to float as a glass cushion, which in turn is supported in an intricate network mesh around the existing dome.

RIGHT AND OPPOSITE: Historic dome and new glass and steel roof.

CONZETT BRONZINI | GARTMANN

Jürg Conzett, of Conzett Bronzini Gartmann (CBG), often works at an engineer's drafting table, sketching in pencil in his office at the center of Chur, a mountain town in Graubünden, Switzerland. In the tradition of Swiss engineer Robert Maillart (1872–1940), who epitomizes the art of the structural engineer,[1] Conzett Bronzini Gartmann (CBG) design structures both in the role of primary designer, such as for bridges, and in collaboration with architects, primarily from the region. The firm's strong local orientation and attachment to the topography, geography, and vernacular building fabric of the surrounding alpine communities as well as their construction expertise equate to an engineer's "critical regionalism."[2] Certainly, the work of other contemporary Swiss engineers also has a regional character—such as Walter Bieler's wooden pedestrian bridge across Lake Zurich, which creates an elongated passage that lightly skims the water, or Christian Menn's highway bridges, which dramatically span deep ravines on their way up mountain passes. CBG's subtle innovations and minute maneuvers imbue their structures with a unique aesthetic that, although it might seem obvious, has wide-ranging technical and aesthetic implications. Theirs is not a practice of emergent materials but of established materials and technologies engineered anew.

Jürg Conzett, Gianfranco Bronzini, and Patrick Gartmann started their firm in 1999, each bringing to the practice architecture and engineering backgrounds and experience gained from working at local engineering offices. Conzett studied engineering at EPF (École Polytechnique Fédérale/Federal Polytechnical School), Lausanne, and ETH (Eidgenössische Technische Hochschule/Swiss Federal Institute of Technology), Zürich, and then worked as a designer in the office of Peter Zumthor from 1981 to 1987. He went out on his own in 1988, started his first practice in 1992 with A. Branger, and then co-founded CBG.

Conzett says that in the 1970s, when he attended university, engineers were not encouraged to discuss design; form was not even an ingredient in the study of structure. But his own focus on design set in motion an exploration of structures that looked beyond just their support systems to the cultural implications of building, environment, material, and shaping space for people. He believes that basic engineering education even today does not provide the problem-solving options and potential range of solutions that a broader, combined study with architecture does. "If you start with a pure formal concept, that means you start with metaphor; you can start with anything, but ultimately it needs to make sense at many different levels. If you ask, why is it made like this? The answer 'Because it looks like a fish' is not a good reason. You should have five answers to that question from all angles, structural, to programmatic, to aesthetic."

Conzett is not interested in engineering feats that result in a new prowess or forceful structural potential. As he says, "Structural engineering makes a lot of nonsense possible, and it is not all on the same level." One of his basic tenets is not to make something just because it is possible; that challenge does not interest him. "It is one way of working, but in our work we are more involved with the architect."

With so many ways to initiate the design of a structure, Conzett says, "Normally you start with intuition, but you have to finish with a rational design." Conzett began the design of the Suransuns Footbridge with freeform sketches. But he says that after it was developed, "I can tell you a logical sequence of thoughts that led to its form. It makes a lot of sense, and I am happy about it. In a similarly rational process you always have decisions to make. But first you have to find the right form for the right category; graphic statics are often more helpful than the computer programs to calculate the forces. To find the form in the old-fashioned graphic statics method of drawing with mechanical tools often is more powerful and faster; you see much more. It is the opposite of free form, it is looking for constant forces similar to nineteenth-century engineers such as Carl Culmann, it is economical."

Conzett also became fascinated with the perception of three dimensions from various viewpoints through the influence of Eduard Imhof (1895–1986), who was an illustrative cartographer, teacher of Conzett's cartographer father, and a longtime teacher at the ETH in Zurich.[3] Imhof sought to create a sensation from drawing with colors and investigated ways to represent layers of depth in order to perceive two dimensions as three. Conzett absorbed this way of seeing and representation as a way to experience existing structures and then design new ones in the landscape. A certain aesthetic tradition of purity in Conzett's work identifies his form-making and his details in structures such as bridges that he designs independently of architects. He notes that "often, engineers make beautiful bridges because they are forced by technical constraints to come to a form that makes sense. This in turn makes the detail a relationship to each specific situation. So I formulate some rules. If you make a retaining wall and the crown of the wall is horizontal, I try to make it extended so that it is parallel to the ridge it is holding. I like to make rules to see what it produces. There are a lot of things that building technique does not influence; you cannot build only based on structure, there are decisions to be made in terms of aesthetics." These aesthetics can be as simple as where the lines of the formwork fall on the concrete and the pattern that results.

In designing bridges, the firm has an opportunity to follow in the construction traditions of their region. In 1996 Conzett, with Branger, designed the Traversina Footbridge, which was suspended between two rock abutments on an old Roman road but was destroyed by a rockslide in 1999. A prefabricated lightweight parabolic, inverted bow arch wood-truss structure, the bridge spanned 47 meters with an incline of 6

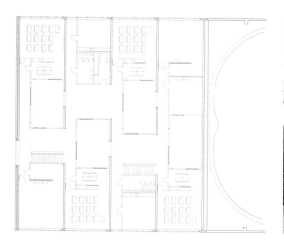

ABOVE: Volta School, Miller & Maranta Architekten, Basel, Switzerland, 2000.

OPPOSITE: Wohnüberbauung Schwarzpark, Miller & Maranta Architekten, Basel, Switzerland, 2003.

PAGE 80: Traversina II Footbridge, Viamala Ecomuseum, Graubunden, Switzerland, 2005.

percent. A light substructure had a three-chord truss girder with two parabolic cables below with a heavier superstructure of a three-ply panel, stiffened girder, which doubled as side rails attached to a horizontal Glu-laminated beam. Vertical struts connected the two structures, and the whole substructure was flown in by helicopter and installed on site. After the bridge was destroyed, Conzett's office designed a second Traversina Footbridge as a series of steps suspended tenaciously between the cliffs, experimenting with structural concepts.

CBG has also been involved in restoring numerous bridges and historic structures, such as the restructuring in stone, wood, and concrete of the Glennerbrucke (2002), which is a hybrid of a new and old structure. And in the historic barracks at St. Luzisteig, with architects Jüngling and Hagmann (2005), Conzett restored the historic elements while integrating a new structure in a collage of existing and new materials and support systems.

In architectural projects, Conzett works closely with architects to develop spatial solutions for the proposed programs. As an outgrowth of the prevailing collaborative practice in Switzerland, one association—Schweizer Ingenieur und Architekt (Swiss Association of Engineers and Architects)—serves both professions, and their publication, *Tec21*, contains articles about both professions, exposing each to the other's processes and techniques. CBG works primarily with regional architects, mostly those from German-speaking Switzerland, such as Gion Caminada, on the School in Duvin (1995); Conradin Clavuot, on St. Peter School, in St. Peter (1999); Jüngling and Hagmann, on buildings such as Ottoplatz, in Chur (1996–98) and the St. Luzisteig army residences (1998–2000), in Maienfeld; Meili, Peter Architects, on the Wood Technology School, in Biel (1991–98); and Miller & Maranta, on the Aarau Market, in Aarau (2003). The firm's only projects outside Switzerland's borders thus far have been the Swiss Pavilion for the Hannover Expo (Peter Zumthor, 2000), the Canale Coupure pedestrian bridge, in Brugge, Belgium (2002), and the Murano footbridge in Austria (1995). Most often he works with architects right from the start and "then there are different intensities of collaboration depending on the knowledge of both parties. If people like each other and have some common values, that helps." This is seen specifically with Dieter Jüngling and Andreas Hagmann, Chur architects with whom Conzett has worked on numerous projects where each critique results in a dynamic exchange of ideas. As Frank Barkow says, "Jürg has the ability to look at the world around him and to distill his observations in often astonishingly fresh ways. In an alpine context like Switzerland you are surrounded by very physical forms of construction: bridge-building, retaining walls, tunnels and roads. The stuff that an artist like Donald Judd used to talk about as the very best of American 'architecture.' In this way Jürg's knowledge oscillates between the historically significant and the everyday. For us this meant that we could work with a structural engineer well-armed and inventive; one who can enable our work in a material and spatially compelling way."

One of Conzett's explorations is how a structure supports a space or encloses it, but in fluid or interlocking ways, not in rigid 90-degree angles—a balancing act. For example, Conzett cantilevered the main volumes of the Trumpf buildings, in Grüsch, Switzerland (Barkow Leibinger, 2001) beyond their foundations, dramatizing the available space and embracing the outdoor space. In the Volta School (Miller & Maranta, 2000), CBG's system of major longitudinal load-bearing walls act as simple parallel beams with the floor slabs, allowing the classrooms to span freely above the 28-meter-long gymnasium that is housed inside a former oil tank trough 6 meters below

ground; the non-load-bearing facades brace the building in the horizontal direction. The horizontal restraint of the floor slab—achieved in most buildings by vertical structural core elements, such as elevators and stairs—keeps the walls in place. The floors are only 250 millimeters thick and are held in place by means of post-tensioning of the cables in the structural slabs in the floor. Staggered courtyards bring light into the 40-meter-deep building and organize the spaces.

The structure of Untertor, an office building for the town of Chur and a television media company (Staufer & Hasler Architects, 2005), includes very few columns. The main facade is a cantilever in the corner by the other facade, staggering the vertical slab corners, which Conzett developed in response to the building's urban site. As Conzett casually remarks, "It is just a little bit of concrete and that is all. Adding little things, we can make big spaces on the ground floor, which is something innovative that you can do, even when you have a small budget."

The suburban train station in Worb, designed in 2003 with the Bern-based architecture firm Smarch, is a project in collaboration with partner Patrick Gartmann. The structure combines the hybrid program of a train station and depot, rooftop parking deck, and bike parking in a stainless-steel banded wrapping that follows the flow of the tracks. The walls are made of one-millimeter-thick stainless-steel ribbons, assembled by steel workers from the region one band at a time; each was pulled tight and then clamped at points between the structural concrete to create an undulating form. Light and shadow animate the structure at different times of day. The cladding is a response to Semper's theories that question what is a wall, in which form follows the construction.[4]

Regionalism is also expressed in CBG's work through the materials the firm selects. For a new footbridge in Vals (the town where the granite baths designed by Peter Zumthor are situated), Conzett is using the same local granite, which the town preferred because it suited the character of the area, even though it was not economical. Wood is the material that Conzett finds the most rich with potential, but it has gone in and out of fashion in Switzerland partially due to an association with wartime rationing of steel. Wood, says Conzett, is never investigated thoroughly for its structural potential by engineers. Because wood is an organic material it moves and causes differential settlement. Conzett nonetheless has exploited wood with new structural technologies, beginning with the Hannover Swiss Expo building, the Wood Technical School in Bern (Peter Zumthor, 2000), the covered wooden footbridge in Mursteg, Murau, Austria (1995, both Meili, Peter Architects), and the school in Duvin. At Duvin one senses that Conzett relished the load-bearing, 12-centimeter-thick larch blocks, using them to create a new building with an age-old system of massive walls and prestressed wooden girders. With the St. Peter School the log structure allows for settlement. The use of wood goes beyond the fascination with regionalism as an exploration for the potentials of organic structures.

In 1996 Conzett participated in a competition with architects Paola Maranta and Quintus Miller for a marketplace, Färberplatz, in the center of the town of Aarau. The project was completed in 2003. This new public space in the city for open markets and events is a sculpted, geometric space that speaks to the theme of pragmatism and simplicity reinforced by Conzett's subtle, nonconforming use of wood and concrete. The main facade defines the dimension of the square building, and is a modified and modern interpretation of a medieval half-timber structure with a solid concrete base and open upper level

with Douglas fir wooden slats. A wooden frame, supported by a central laminated timber column with four beams that radiate outward to support the roof, is connected to smaller, lamellae-like angled wood elements. Using computer numerical control (CNC) milling technologies, each piece is set at a different angle, creating a nonorthogonal interior but a regular ceiling. Concrete walls at occasional points stiffen corners and protect the building from displacement and movement.[5]

Many engineering projects are completed only through the design phase in a competition; however, these unbuilt projects still contribute to engineering culture, through the dialogues that go on between engineer and architect. For example, Hardturm Stadium, in Zurich, (Meili, Peter Architects) whose construction has been delayed, was conceived as a multiuse facility, including hotels, cultural activities, a convention center, and residential units. Its pentagonal shape emerged with Conzett's input on how to structure the form so all of the programmatic requirements could be accommodated within the tight urban site. One of Conzett's main design contributions to the stadium was in structuring an open central podium, supported by truss girders that became oblique supports beneath the grandstand, which helped to increase the horizontal flexural and torsional resistance.[6]

CBG's approach to projects is to identify how structural elements can solve many problems simultaneously. Conzett says, "It is just a method in which to work, but it leads also to a certain aesthetic, in an exchange of ideas."

OPPOSITE: Worb train station, Smarch, Bern, Switzerland, 2003.

ABOVE: Farberplatz, Aarau, Miller & Maranta Architekten, Aarau, Switzerland, 2003.

SURANSUNS FOOTBRIDGE
VIAMALA, SWITZERLAND

Conzett Bronzini Gartmann has designed small footbridges that reveal the poetics of their forms and structure and reflect the firm's dedication to the specific characteristics of material, methodology of construction, and constraints of site. In one region of the Graubünden, on the historic Via Mala, in the largest gorge between Milan, Italy and Constance, Germany, CBG designed the Suransuns Footbridge (1999), a delicate stone suspension footbridge. Slung across the Hinterrhein, it resembles a minimal Himalayan rope bridge and leads hikers through the outdoor Ecomuseum of the KulturRaum.

Conzett resurrects systems of prestressed stonework and concrete normally used in arches, also seen in Jörg Schlaich's stressed ribbon footbridges over the River Enz (1992). CBG applied a traditional system to a new structure, using a combination of materials that could withstand the pressure and weight. In this prestressed structural system—like the 1954 project for a granite bridge, Teufelsbrucke, designed by Swiss engineer Heinz Hossdorf on the Gottardstrasse, but here in prestressed stone and steel—the stainless steel ribbons are sandwiched between transverse steel plates and the granite footpath slabs, which are held in place with aluminum at the joints. The two parallel steel ribbons run like tracks beneath the locally quarried, 6-centimeters-thick Andeer granite treads. The ribbons terminate in tensioners anchored into concrete abutments. The stress ribbons and the pavers act together as a single suspension system, and the stiffness and weight of the stone resist swaying. Primarily, it is an upside-down arch in a slender, reduced form. Minimalist stainless-steel vertical rods tie the steel plates to the granite slabs and to the thin stainless-steel handrail above. The design of the bridge engages walkers, who at their own pace can focus on what it means to traverse a narrow area in a weightless and suspended state.

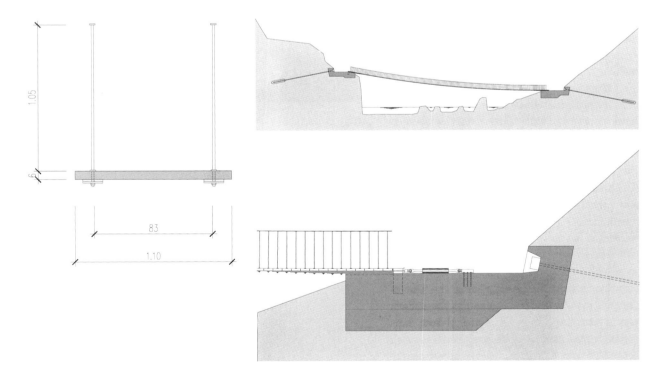

CLOCKWISE FROM LEFT: Detail of steel rails; section showing anchoring system; section showing platform and stays.

OPPOSITE: Andeer granite treads and connections (above) and view across bridge (below).

CONZETT BRONZINI GARTMANN | 87

TRAVERSINA STEG II

VIAMALA, SWITZERLAND

Conzett Bronzini Gartmann completed a new bridge in 2005 to replace their first Traversina Steg, a suspended wooden footbridge, in Graubünden, Switzerland, sponsored by the Viamala Ecomuseum, which was destroyed in 1999 by a rockslide. The location—high over an impassible 100-meter-wide ravine—presented the challenge of building a bridge that was functional, beautiful, and appropriate to the national park setting. The bridge was to link the old Roman trail through the rocky and wooded area and emerge from nature, not overwhelm it.

Conzett began the bridge design with technical hand drawings using graphic static calculation methods, which were then transferred to computer-based form-finding calculations. The new suspension bridge is seemingly precariously placed across the ravine, and has a structural lightness similar to that of the nearby Suransuns Footbridge. During the design stage, the engineers developed four variations: a stair as a tension band, a suspended stair on a cable framework, a suspended podium stair, and a bridge below. Ultimately, the suspended stair on a galvanized steel cable framework was selected, with pylons made using the rock of the valley supporting the bridge. Two large concrete abutment pillars support 100-meter-long steel cables that crisscross in a post-tensioned, wire-cable framework, as a diamond truss system organized in two vertical planes to support the larch walkway, which is also kept rigid with prestressed main cables. Structurally it is similar to a Jawerth truss system (an unusual truss system by Swedish engineer David Jawerth where two suspension cables are stressed against each other using a zigzag of struts) but the lower cables form an upside-down bow, rather than being draped from pylons. This crossing, not suited for those with vertigo, consists of a series of 176 wooden steps in a 62-meter-long hanging staircase forming the better part of the 95-meter-long bridge that drops 22 meters from the north to the south. When Conzett takes visitors to the bridge, he insists on the approach from the north to have the full effect of the view and the dramatic stairs. The engineers considered the comfort of the hiker in the design, in terms of what they could see and feel, so the space between the treads is minimal because of the angle of the shelflike stair, balustrades of slatted wood were used instead of thin cables, and the slope has a walkable incline. The structural effect makes hikers feel as though they are in a protected trough.

The challenges of the environment and varying weather conditions, such as wind and snow loads, were key in the conceptual development of this simple system. Water can run off, and snow does not accumulate on the surface. Elements can also be replaced easily, facilitating maintenance. The cables terminate at the top of the pylons to simplify their post-tensioning. Similar to Suransuns, both the handrails and the abutments are minimal in design; this purity of structural expression relates to the natural landscape setting in which the bridge was inserted.

RIGHT: Tread detail section.

OPPOSITE: Section showing the main anchoring points (above); bridge in landscape; view from north to south; opening day (below).

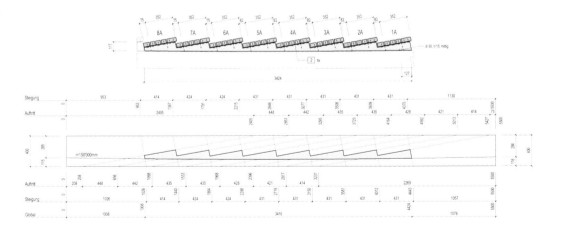

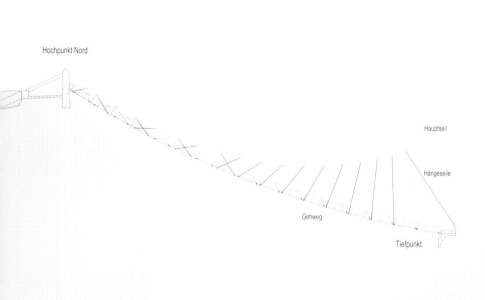

TRUMPF PAVILIONS
GRÜSCH, SWITZERLAND

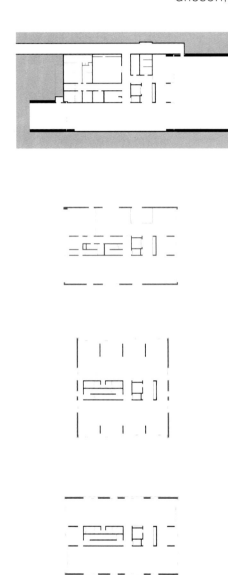

Conzett designed the structure for two buildings with Berlin architects Barkow Leibinger for the Stuttgart-based laser machinery company Trumpf, near the village of Grüsch, in Graubünden, Switzerland. One structure, a center for exploring ideas in manufacturing technologies and metal fabrication, was completed in 2001, and the other, an administration building completed in 2004, are each connected separately by tunnels to the 1970s machine-tool factory. The manufacturer was forward thinking in siting research and development near the production facility, a strategy that also provides opportunities for start-up companies. The structures for both buildings are demonstrations of the engineer's repertoire of bridges, retaining walls, excavations, tunnels, and cantilevers applied to architecture. The two buildings are each stacked, rotated volumes made up of concrete slabs that form a wall-on-floor system, but in a continuous-pour. The building was shored up with logs and when the tensioning was set with the concrete, the shoring was knocked out and construction continued. The concrete is intensely structured with rebar. The exposed concrete at the lower two floors and interior upper levels is either formed with smooth offset formwork panels or, as on the retaining walls, is washed to expose the aggregate. In contrast in the cafeteria, a terrazzo floor provides a smooth surface, resulting in three textures describing the solids.

Frank Barkow emphasizes that in their collaboration Conzett designed a structure that created enough depth for programmable functional space, forming a dialectical relationship with the landscape that enhances the architectural effect. The stacked arrangement of the building blocks pragmatically protects entrance spaces under the cantilevers, which also form garden terraces above, and opens up uninterrupted floor space for the workshops and training areas. The supporting structure is made by stacking concrete walls, planes, and floors in a monolithic way that resembles hollow box beams. This principle is economical and efficient in the sense that it utilizes exterior and core walls and floors for support, eliminating the need for interior structural supports, notes Conzett. The main entrance, reached via a ramp or a series of stairs, leads to the lobby and the main workshops on the north–south axis, and bridges over a depressed landscape that slopes beneath the building to the cafeteria. On the lower level, this bridging creates a space between the retaining walls for the cafeteria and the entranceway to the tunnel. Simultaneously,

ABOVE: Pavilion I showing the variation in the elevation.

BELOW: Pavilion I showing cut into landscape and concrete retaining wall.

OPPOSITE: Plans of the wall systems at each level of Pavilion I.

CONZETT BRONZINI GARTMANN | 91

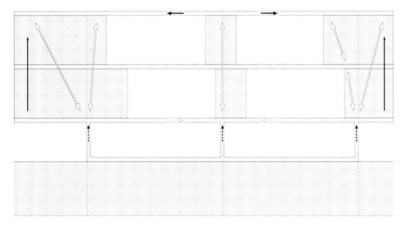
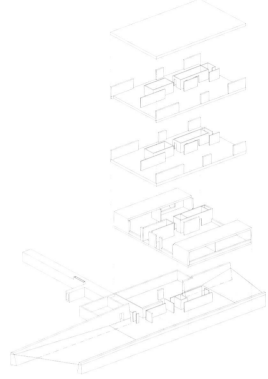

this recessed ground allows the height of the building to be reduced, in accordance with zoning regulations. Stairs and elevators lead to the office space on the two upper floors, which are differentiated on the exterior with a red-stained larch cladding.

The second building, a research center for laser cutting machines is, like the first, situated diagonally on the site and is formed by three rotated volumes bridging a cut in the ground and running parallel to the valley. The 3,480-square-meter space houses workshops, laboratories, production areas, and distribution; in essence, it operates as a small factory. Offices and conference rooms on the north side, and open office areas on the south, are reached by way of a central service core with a double stair around which the building is organized. Workshops and the factory spaces in the basement receive natural daylight beneath an excavated ramp to the east. The structural expression of the building is essential to the form and composition, embodying the architect's intent.

ABOVE: Sectional diagram with the shearwalls and exploded axonometric showing floors and the shearwalls (above); construction with rebar in place (left).

OPPOSITE: Wall section showing cantilevered structure.

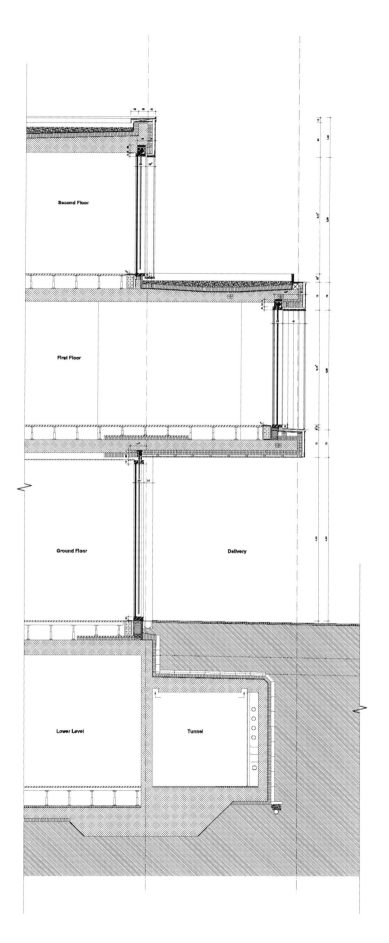

DEWHURST MACFARLANE AND PARTNERS

The performance of glass as a building element has surged progressively from the small pieces used in the eleventh century for leaded and stained glass windows to the immense sheets used for twenty-first-century curtain-walled skyscrapers. The experimentation and innovation occurring with glass today is similar to that of concrete and steel in the early twentieth century by engineer-builders, often called *constructeurs*, such as Gustav Eiffel or Jean Prouvé, who invented structural systems in coordination with new industrial products. Today it is the work of the engineer, the architect, and manufacturers in collaboration that pushes glass, or any material, to its fullest capability. The engineering firm Dewhurst Macfarlane and Partners provides both structural and facade engineering services for all types of materials and forms, but has become most recognized for their innovations in glass—looking at the "design intelligence" behind the material and its functional process and considering ways to transform a nonstructural material into a structural one. The London-based firm was founded in 1985 by Tim Macfarlane and Laurence Dewhurst and has grown to have a staff of thirty between offices there and in New York City.

Whether addressing an architect's project or initiating his own structural designs, Tim Macfarlane begins a project with a blank sheet of paper, signifying the open mind he brings to the specific problem at hand, without any preconceptions or generic solutions. Macfarlane treats each structural situation as unique, seeking to repress standard responses. In teaching structures to architects at the Architectural Association, London; Illinois Institute of Technology, Chicago; and Yale University, New Haven, Connecticut, he also focuses on the integration between architecture and engineering and similarly encourages students to focus on process in their design solutions.

Macfarlane creates a similar dialogue with architects, appreciating the architect's design and intentions. Searching for a definition of function, which is also performative, he searches to express the relationship to the social and industrial circumstances from which a design is based. Macfarlane directs his analytical energies into static structures, rather than industrial design or mechanical objects. He compares engineers to nomads or hunter-gatherers who practice their art with an almost monotheistic focus. Engineers are usually trained to identify "the" problem and hone in on "the" solution.

He emphasizes that "a good example of a pure engineering problem could be the design of a spacecraft where form will definitely be subservient to function. It is unlikely that in engineering a spacecraft the engineer would be inclined to equip it with anything more than the absolute minimum necessary to sustain life and deliver the craft safely to its destination." Macfarlane continues, "All forms of transport share this fundamental core challenge of managing dynamic forces in an efficient way, although it is ultimately difficult to suppress the impulse to decorate and embellish once the forces have been adequately 'managed.' Deleuze and Guattari's essay '1227: Treatise on Nomadology—The War Machine'[1] considers the engineering impulse to create war machines or weaponry as part and parcel of a hunter-gatherer nom(mon)adic existence. Bruce Chatwin in *Song Lines* paints the image of Moses leaving the flesh pots of Egypt, with its multicultural Babylonian richness for the barren desert with ten clear rules and one God."

Architects on the other hand, he suggests, are urbanists whose role in translating the cultural aspirations of a society into built form demands a broader polytheistic or secular frame of mind. Cities are layered structures that have grown organically while reflecting their cultural complexity. He notes: "This accretional development encourages the slow 'static' arrangement of structures to create a condition or 'state' of equilibrium." The engineer's role in city-making is therefore more broadly focused than in machine building, requiring a different set of skills and level of interest.

As an engineer interested in concepts of nomadism and individualism, he twists the norms of material properties to make what is static extraordinary. For him, "dynamism is not urban, it is superurban. There is the thrust and movement in the city, but it is still."

Macfarlane's interest in static structural compositions and the balance of structural elements drives the design aspects of his work, similar to an architect's interest in light and shade, and he blurs the dividing line between compositional and structural design. Macfarlane intuitively visualizes the bigger gestures and how a structure engages space. He wants the architect's concept to resonate, broadening what structure can do and enhancing the overall design. The engineer becomes the facilitator for the design's optimization, applying formal rigor to the solutions and helping to bring the architecture into existence. Macfarlane prefers solving multidimensional programs with architects to distill the nature of a design and transform the notion of authorship in a nonhierarchical process.

ABOVE: Now and Zen Restaurant, Rick Mather, London, 1991 (left); House addition, Rick Mather, Hampstead, North London, 1992 (right).

OPPOSITE: Yurakucho subway station, Rafael Viñoly Architects, Tokyo, 1996.

PAGE 94: Staircase of Ebury Street, Joseph Shop, Eva Jiricna, London, 1994.

Peter Rice's structural glass systems were a jumping-off point for Dewhurst Macfarlane's projects. The glass walls Rice developed in the 1980s were supported in a steel framework with a system of complex fittings, where the stresses were carried by the supports, not by the glass. A chief characteristic of his work was the tensegrity, or tensional integrity, of his structures in which the structure is formed by tensional rather than compressional systems.[2] Rice translated these concepts into manufactured systems featuring steel clamps or bolts in point supports and cable suspension systems that supported the glass.

In a departure from Rice, however, Macfarlane is interested in glass performing the structural work independently, without the structural intensity and clutter of cables and frameworks. This goes hand in hand with his interest in specialized projects, achieved through computer technologies and design integration to customize cladding. Structural and architectural codes are fixed in building manuals; progress in building practice is often in the hands of industry. Companies such as Pilkington held the patents on flat plate-glass systems for limited periods, and Macfarlane's investigations into structural glass occurred simultaneously with the industry's need to expand. This new potential gives Macfarlane a "creative rush that is infrequent in engineering. Glass is a difficult and often unpredictable material, psychologically and physically: It shatters; it cuts; it is dangerous." But he incorporates the complex properties of glass in his projects, taking it further than expected or previously known.

One can trace a chronology of later-twentieth-century innovations in structural glass within the Dewhurst Macfarlane firm through a series of projects. The baroque expression articulated in the connections of the glass staircases designed for the Joseph shops in London (Eva Jiricna, 1994, 2003) became an opportunity for the architect and engineer to put glass to use structurally. In order to achieve the required strength for the Sloane Street location, the treads needed to be 19-millimeter-thick annealed glass[3] covered with a heavy, 15-millimeter-thick sheet of acrylic that attracted dust; the design aesthetic was less than appealing. Macfarlane proposed laminating glass instead, using acrylics as a sandwich to augment its performance and make the material composite stronger and seamless. This material was used successfully in this first application, and it was inserted into the lighted flooring in front of the Now & Zen restaurant on St. Martin's Lane, also in London (Rick Mather, 1991). In that project, the engineers specified 3.6-by-.9-meter annealed glass in much thinner sheets than in previous projects, laminated with acrylic resin with four-sided support. The annealing ensured stabilization in these large glass plates. For the house addition in Hampstead, North London (Rick Mather, 1992), Dewhurst Macfarlane used a triple-laminated annealed glass in 9.5-millimeter-thick sheets to form the columns and beams of the portal structure. The beams took the shape of fins, and the components were attached with mortise-and-tenon joints and a structural silicone, deriving additional support from the existing building's wall. This innovative method became Macfarlane's signature.

Perhaps the firm's most pivotal project using structural glass was for the Yurakucho subway station for the Tokyo International Forum Plaza, in Tokyo (1996). Macfarlane designed, with the architects Rafael Viñoly & Partners, a cantilevered glass canopy 5 meters wide, spanning 11 meters. As a load-bearing structure, three parallel laminated

glass sheets toughened with Plexiglas panes are cantilevered up and over the station entrance. The 19-millimeter glass sheets were then bolted together at connections, which transferred the forces to prevent high stress that would otherwise lead to failure.

The firm applied the knowledge-base gained from the canopy project to their work on the structural glass facade and 10-meter-long beams for the Samsung Jong-Ro building in Seoul, Korea, and to the Kimmel Center for the Performing Arts, in Philadelphia (both Rafael Viñoly Architects, 1999, 2001). The 1.5-by-1.2-meter laminated glass panels forming the Kimmel Center's 55-meter semicircular wall were supported by single-direction, parallel cables. This model project has resulted in a new structural code for laminated glass, and the firm put it to use in other projects, such as the Lewis-Sigler Institute for Integrative Genomics at Princeton University (Rafael Viñoly, 2002) and the addition to the Marion County Library in Indianapolis (Woollen Molzan and Partners, 2005). While the geometries of the facades varied, they used a similar tensioning system with weights tensioning between the roof structure and foundations. The challenges in each case were to stop any warping of the glass panels, thus ensuring that the cables be maintained as designed, and address wind loads. For these projects, the bidding contractors' lack of familiarity with two-way cable walls meant Dewhurst Macfarlane remained involved in the construction phase.

Strength and transparency are the qualities Macfarlane seeks in the glass materials he selects. The laminated glass the firm developed for the treads of the central 2.4-meter staircases in Apple Computer's stores in Los Angeles, Chicago, New York, and San Francisco (Bohlin Cywinski Jackson Architects, 2002) conveys a sensation of lightness and luminosity, while being strong. SentryGlas Plus, a DuPont lamination product, was used to secure the treads inside the layers, reinforcing the glass in a manner similar to reinforced concrete, creating both a solid and transparent structure. For the unbuilt air-traffic control tower at the Barcelona Airport, controllers needed an unobstructed view to monitor airplane traffic. The window glass would be held in place at its edges with only a structural pin and then fixed in a channel so that the inflection of the one would react against the deflection of the other, minimizing bowing. In an addition to the Children's Hospital in Philadelphia (Kohn Pedersen Fox, 2006), the architects had specified

a glass entrance wall. Rather than use the more standard glass coated with Teflon in a monolithic sheet, Dewhurst Macfarlane proposed a simplified system of large panels of toughened laminated SentryGlas Plus. Working together, the engineers prompted the glass manufacturers to adopt a new lamination method, and manufacturers Pilkington and DuPont are developing a new product, the SentryGlas Plus laminated planar system. This represents a significant shift in the industry standards for this type of laminated glazing.[4]

Dewhurst Macfarlane also engages in non glass-oriented projects such as the Bengt Sjostrom Starlight Theatre, at Rock Valley College, Rockford, Illinois, an open-air community theater (Studio Gang/O'Donnell, 2003–5). This limited-budget project involved expanding the existing 600-seat amphitheater, adding a fly bridge, extending the entrance pavilion, and devising an operable roof system that could open and close according to the weather conditions. The benefactor, Sjostrom & Sons, was also the contractor for the project, which had begun in 2001 with the addition of round windows arranged in a constellation-like pattern in the 5.5-meter-high concrete rear structure. Macfarlane was involved from 2002 to help design the 15.25-meter-tall, copper-clad fly bridge for over the stage, and the sliding, translucent doors, similar to airplane hangar doors, that were to substitute for curtains to create an additional rehearsal space.

Macfarlane believes that design intelligence is formed as an engineer works on an architectural project, where the innovation and process in attention to design is different from the manufacturing of an industrial product. "In the architectural realm you don't get the opportunity to research something before you build it. The projects become the research. Design intelligence expands an area of knowledge, not only the creation of a specific element. It is an invention of a complex system." How the material is repurposed in the architectural world expands the core of knowledge and design, which to him is the "tipping point." The techniques that further design intelligence in engineering suggest how to represent a shape technologically and produce it. Macfarlane emphasizes, "Each manipulation of a form, in physical reality, can defy gravity. Then the question must be asked, Is it reasonable to build?"

ABOVE: Philadelphia Children's Hospital addition, Kohn Pedersen Fox, Philadelphia, Pennsylvania, 2006 (left); Bengt Starlight Theatre, Studio Gang/O'Donnell, Rockford, Illinois, 2003–5 (right).

OPPOSITE: Apple Store, Bohlin Cywinski Jackson Architects, New York, 2006.

KIMMEL CENTER FOR THE PERFORMING ARTS
PHILADELPHIA, PENNSYLVANIA

Dewhurst Macfarlane and Partners were the lead engineers for the Kimmel Center complex in Philadelphia, designed by Rafael Viñoly Architects in 2001. Two concert halls and light-filled public areas under a 130,000-square-foot glass barrel-vault roof with glazed end walls create a vibrant cultural gathering place for the city. Tim Macfarlane, in recognizing that a slender, stiff structure would require a folded plate truss, designed a barrel vault to give the basic form to the roof. Six-foot-deep Vierendeel trusses sloped at 60 degrees form the ribs of the vault, creating a folded-plate structure in cross section. This structure also accommodates conventional rectangular glass panels. The Vierendeel truss throws the loads back down to the supports at the sides without the need for cross bracing, keeping the center vault open and maintaining a uniform and simple shape. The zigzag herringbone pattern of the roof form recalls the historic Palm House (H. & D. Baily, 1843) in Bicton, England.

The two 12,000-square-foot glass end walls were designed to have a cable structure supporting low-iron, fully tempered laminated glass panels measuring 4 feet 1 inch-by-5 feet 9 inches, similar to Jörg Schlaich's Kempinski Hotel in Munich (Murphy/Jahn, 1994) but in an arch form. However, rather than use the tennis-racket-type cable net in two directions, the engineers supported the cables tied to one separate arch. The 185-foot span called for stability from the narrow cables, but instead of an enormous truss, the engineers intuitively proposed only vertical cables. In order to maintain tension, vertical cables are hung from the arch and held in tension by cast-iron weights at the base of the building. Similar to the free-form flow of the ideal catenary arch, the weights are constant, but the cables are flexible in the wind and are

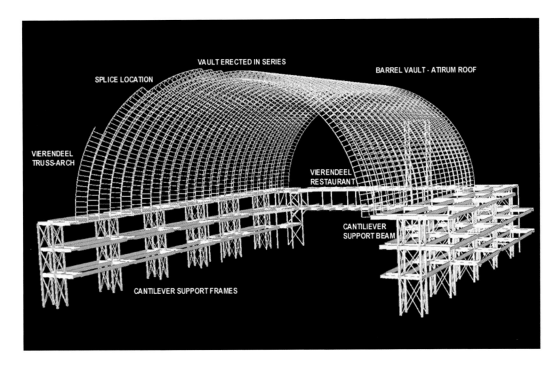

ABOVE: Structural diagram of main volume.
OPPOSITE: East facade showing cable structure glass wall.

ABOVE: Aerial view and detail of roof pattern.
OPPOSITE: Main entrance on Broad Street.

designed to deflect up to three feet in the center of the double-channel steel arch, which is distinct from the barrel vault. Macfarlane discussed the issues with specialists such as engineer Michael Barnes, director of structural and lightweight engineering at the University of Bath, as well as with other peers, and mock-up tests confirmed the structural stability of the proposed system.

The construction manager wanted to contract the work to a design-build firm so that the responsibility for the glazing system would be with the contractor, but the estimates came in too high. Macfarlane explained to a local glass company how the roof would be built and the company agreed to take it on. "Kimmel illustrates more than other projects what moves beyond in an engineering sense to me, and what it is that we do to push a discipline." Macfarlane feels generally that engineers themselves "should do design-build," because "if we are really contributing to the process we should take responsibility and instruct the builder how to proceed. But most engineers don't want that responsibility. It is not good for the industry and innovation as a whole. The contractor won't take a risk if he hasn't seen it before. When you are working with materials from the 1930s there isn't much room for innovation . . . Glass can move that boundary," Macfarlane concludes.

DAVIES ALPINE HOUSE, ROYAL BOTANIC GARDENS

KEW, LONDON

The replacement Davies Alpine House at the Royal Botanic Gardens, Kew, designed by architects Wilkinson Eyre and completed in spring 2006, was among the projects recommended in the architect's 2002 Strategic Development Plan for the Royal Botanic Gardens. The existing greenhouse was filled with obtrusive mechanical systems, so the goal for the new structure was to design a lightweight glass house with a concealed air-flow system. Tim Macfarlane began his work on the structural design by engaging in a joint sketching session with the architects.

The building's shape derived from two sections of a cylinder propped up against each other forming parabolic arches where the cylindrical profile of the wall complements the structural elements. Dewhurst Macfarlane's cable systems both at the Kimmel Center and at the Alpine House are single-direction systems. The curved structure of the Alpine House, 144 square meters, raising 10 meters at the central spine, gives it horizontal stiffness, and the glass supported by parallel vertical cables, like a harp, is held down through the joint with pin fittings. The cables follow the horizontal line and create the points to which the 1-by-2-meter glass panels are attached. The structural achievement occurs when it moves out of the flat plane to the curved wall, where the curve in one direction allows the support of the other. Rather than have cables in two directions that would be constrained, because the cable following the curve would pull the other cables, forming a hyperbolic parabola shape rather than the desired curve, the one-way cable was selected. Because of the Alpine House's geometry, the structural framework of the glass is separate from the cables and is not supported in the same way that Macfarlane had previously configured. The glass panels are held at their corners with plate clamps held off the cables on a long pin. The outward horizontal curvature braces the glass against wind loads. The balancing between the two arches made construction tricky, said Macfarlane, but the benefit was that the form allowed for an open-stack ventilation system.[5]

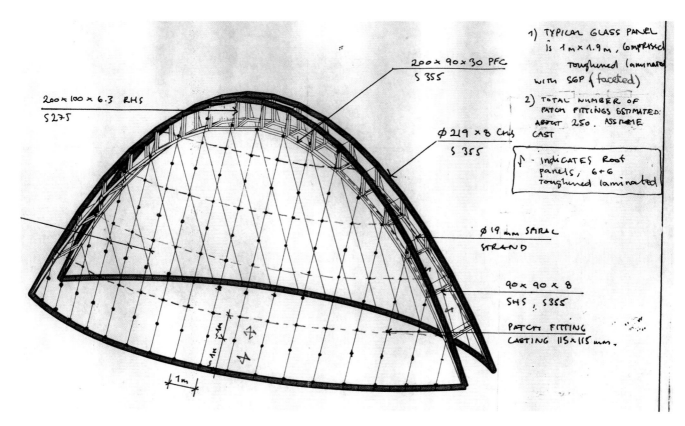

ABOVE: Sketch of structural steel.

OPPOSITE: Model showing shades in three positions.

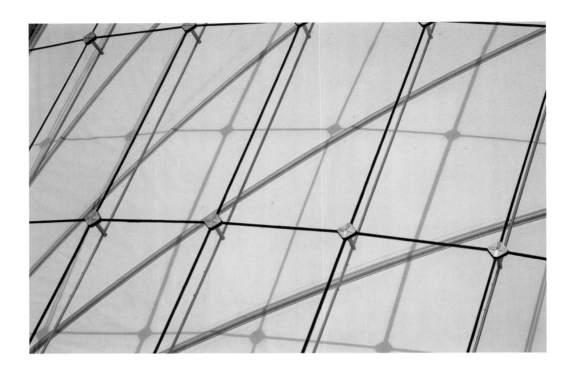

ABOVE: Details of steel connections.

OPPOSITE: Pavilion with shades up and retracted.

The environmental strategy to simulate a cool alpine climate was synthesized through a collaborative process between structural engineering, systems engineering, and architecture. Working with Atelier Ten environmental engineers, the team devised a system for the air flow and ventilation to move through a concealed underground labyrinth of two concrete slabs, which cooled the air and sent it through the center to be delivered through pipes around the edges of the green house to the plants. Custom-manufactured blinds, in the interior, resembling Japanese fans filter out ultraviolet rays and capture the heat. They can be manipulated manually, allowing the building to open and close as needed.

The project demonstrates a set of principles common in Dewhurst Macfarlane's practice, in which the architectural design springs from the integration of performance and form.

EXPEDITION ENGINEERING

The London firm Expedition Engineering was founded in 1999 by structural engineers Chris Wise, Sean Walsh, and Chris Smith (who is no longer with the firm) and includes environmental engineer Ed McCann. In his twenty years with Ove Arup & Partners, and as a director from 1993 to 1998, Wise collaborated with architects such as Richard Rogers and Norman Foster. Wise found Ove Arup's philosophy of "total architecture" to be a liberating one, and it influenced him significantly.[1] When he joined Arup, the staff included two thousand engineers and fifty architects, and the engineers would work on a great variety of projects, from systems engineering to archaeology. Wise explained that "you could do what the hell you liked as long as it was ethical and commercially viable."

But this openness was often difficult and pushed the limits with such projects as the Millennium Bridge in London (Arup/Foster/Caro, 2000), resulting in lessons from failure that Wise turns into knowledge to enhance future projects. After larger-than-expected volumes of opening-day pedestrian activity caused the Millennium Bridge to move laterally more than calculations had indicated it would, the Arup team solved the problems by installing additional damper systems. Dampers that had been included in the original structural design were eliminated after 150 separate analytical and model tests had suggested they were unnecessary. The bridge, conceived on a napkin by Wise and engineer Roger Ridsdill-Smith and developed with Norman Foster along with sculptor Anthony Caro, had to span across a river 300 meters wide. After investigating numerous concepts, the designers came to the idea of the structure as a cable thrown across the river with a flat deck and they became intrigued with the purity of the diagram. Wise feels no reluctance in pointing to the bridge project as a model for other complex projects, both as a cautionary tale for increased testing and as an exemplar for ingenuity in problem solving. And as a

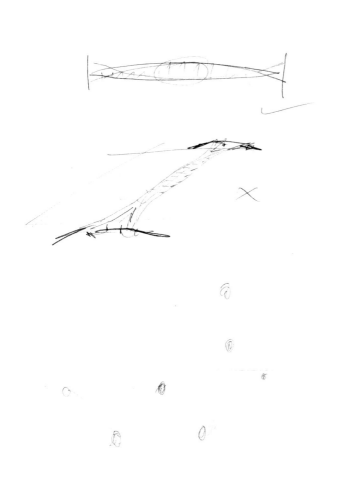
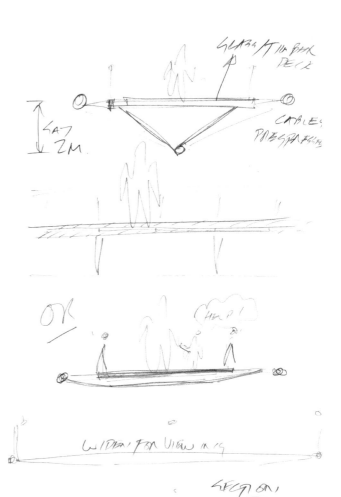
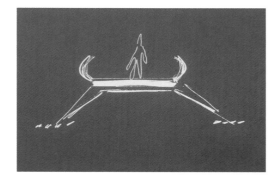

result of the studies done for the modification of the Millennium Bridge, the engineering world has focused on the pedestrian-bridge movement issue generally and improved its knowledge base.[2]

Wise points out that throughout recent history, engineers have been expected "to be infallible, but we make mistakes too. The premise of infallibility pushes us to go backward rather than moving technology forward." Engineering is not a science, so for Wise, "engineering starts with something inside one's head and then is put out into the world; it has a physical product. It is really the opposite of science, which studies what is out there in the world." Wise emphasizes that "you try to come up with something that is on the edge of being possible and then try to prove it." Pushing the profession of engineering to the threshold of technological change while still working within the bounds of safe practice is the aspiration of Wise and his partners at Expedition.

Before the Expedition Engineering team moves on to testing a design in the physical world, the conceptual design process drives their work. They sketch projects in a "brain dump," or a release of ideas onto paper, resolving design issues as structural designers, but as neither solely architects nor engineers. Wise works without a set formula or preconception, and his design work and sketches reveal an underlying logic informed by an intuition similar to what architect Renzo Piano has described as "the turbocharged application of experience," floating in subconsciousness. Wise explains that this experience is one in which, "if you were to build on the process that leads to a sketch . . . you use bits of understanding that you pick up through the years—you can't fake those linkages." His drawing tools are also particular—a soft pencil and thick paper in a large sketchbook, in which most of the projects are expressed in curved lines and fluid forms. He notes, "Design engineering is like having learned a musical instrument: You can just sit down and play; but the notes we use are different from what the architect would use to design a building; what a lawyer would use preparing a case; or what a general would use to fight a war. The whole process is one of having enough confidence to let go to feel what the answer is, and implied by the answer or the diagram are the layers of bits and pieces—that is the thing that makes it so powerful."

The firm strives for efficiency and economies in all of its projects, which Wise tries to achieve by provocatively tweaking standard structural components. He might invent an economical, minimal concrete beam, which he calls "a banker's beam," with a hollowing in the center that removes two-thirds of the beam, stripping out unnecessary material and paring it down to its lightest and most efficient form. He also looks at ways for building components to do more than one job, as in Roman and Gothic architecture, in which the natural light, the structure, the architecture, and environmental systems perform many functions and are beautiful as well. This approach is evident in his work for Arup on the design of the American Air Museum, Duxford, England (Foster, Arup, Roger Preston + Partners, 1997), which in its 90-meter volume formed by two precast-concrete shells, each only 100 millimeters thick, integrates form and function. The flow lines organized the form and the structure, and the environmental and structural elements coalesce. The toroidal geometry is pierced with components cut out from a whole sphere and 924 precast panels made from six sets of standard shell components form the roof, from which aircraft are suspended via secure cast-steel sockets.

While still at Arup, Wise worked on the Commerzbank headquarters, in Frankfurt, Germany (Foster, Arup, Roger Preston + Partners, 1997). His team consulted on the radically new design of a 56-story triangular tower, whose three 16.5-meter-wide office blocks center on a full-height open atrium space that ventilates and brings light into the office areas and the common spaces, among which are newly conceived communal "sky-gardens." Although

ABOVE: American Air Museum, Foster + Partners, and Roger Preston + Partners, Duxford, England, 1997.

OPPOSITE: Sketches and completed Millennium Bridge, Foster + Partners with Anthony Caro, London, 2000.

PAGE 108: Sketches for the North Shore Footbridge, Stockton-on-Tees, England.

 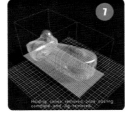 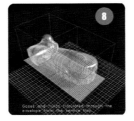  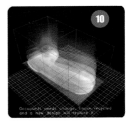

constructed with mass-produced, standard parts, it was a unique, social working environment for various size Commerzbank employee teams who work in twelve-story sub-divided "villages" for 650 people. Wise's structure was instrumental in creating the environmental diagram, ultimately providing a structural design that enabled natural light to penetrate the skyscraper and making an important contribution to the first naturally ventilated tall office building.[3] The structural framework is fundamental to allowing for the twelve enormous openings on the facades that in turn allow natural air and light to flow through the structure. Other structural innovations included the exoskeleton structure that functioned without a central core, which is normally what carries the vertical load and stabilizes a building; a structural exterior comprising a series of load-bearing bridges, similar to the trusses for the Hong Kong Shanghai Bank building; and a standardized, mass-produced construction kit-of-parts that kept the frame to only twenty different components—the schedule fit on one piece of paper—as compared with, for example, the Eiffel Tower, which is an equivalent height but used thousands of unique pieces.

The Torre de Collserola (Robin Partington for Foster + Partners, 1992), a telecommunications tower built in Barcelona, in anticipation of the Olympics, was a groundbreaking project for Wise while he was still with Arup. The 305-meter-tall, 7,432-square-meter tower, which includes thirteen floors of antennae, signal processing equipment, and a public viewing gallery, was raised on one hollow concrete pole 4.5 meters across. Built in one pour at ground level, the 3,500-ton floor unit was jacked up the circular, hollow shaft pole, braced by three vertical steel trusses spaced 120 degrees

apart and steel guy wires in three pairs. The integrated skeletal system carries both the lateral and vertical loads.

Wise has taught at Imperial College, London, and in design studios, specializing in creative design for engineers and engineering for architects often focusing on "future proofing" urban development areas. He has worked intensively on academic curriculum and bringing to light the creative process of scientists and technical experts. Wise also copresented on the 2002 BBC-TV and Discovery Channel program Building the Impossible, a four-part series on early inventions—a submarine, an airship, an Egyptian shaft tomb, and a Roman catapult—that demonstrated the trial-and-error process of inventors.

Expedition, in their broad base of projects, have also noticed a paradigm shift in engineering practice and contract assigning in the industry in general. They note that not only are they involved from the outset in designing a project with an architect, but they are increasingly asked by larger clients to take the lead on vast environmental and other infrastructural projects, such as a tower for the former Peninsular & Oriental Steam Navigation System Company. In cases such as these, they have begun to sit alongside clients to choose architects.

The firm also is exploring new technologies in competitions, such as that in 2005 for the Halley 6 British Antarctic Survey research station (Expedition/Hopkins/Atelier Ten) at the floating Brunt Ice Shelf. Their concept was for a compact, self-sufficient, plug-in building that can withstand temperatures of -50 degrees Celsius and have a minimum impact on the environment. One particular requirement was that the unit be able to relocate. Expedition took this literally and placed the building on automated robotic legs.

In a 2000 competition for the house of the future, the 2020 Concept House (Expedition/Melon), Expedition designed a machine that can make any shape building in a one-print process. It was a proposition that asked, "If you can have everything you want, and you are not inhibited by the manufacturing process, what would you do?" In this building-synthesis machine, rather than coordinate fifteen different manufacturers, numerous pieces, or disjointed interfaces, there would be a single designer/producer going one step beyond the digital fabrication technologies. These projects—which he often calls "mad design"—signify that which is taken beyond the normative to another level of experimentation.

As Wise emphasizes, "In a world where everyone is risk averse, the worst that you can do is to be uncertain when something might not work; pioneering engineering and far-out projects have to bring the client into the adventure." Many engineers argue that just because you can build it does not mean you should. But Wise says, "Just because you can't build it doesn't mean you shouldn't. Or as described in an old Celtic saying: 'The gods show many possible futures. It is up to the living to show how it is done.'"

ABOVE: Proposal for competition for the Halley 6 British Antarctic Survey research station with Hopkins & Partners and Atelier Ten, 2005.

OPPOSITE: 2020 Concept House, Melon Architects, 2005 (above), Torre de Collserola, Robin Partington for Foster + Partners, Barcelona, 1992 (below).

LAS ARENAS
BARCELONA, SPAIN

Expedition Engineering is a key member of the team, with Richard Rogers Partnership, transforming an abandoned brick-and-steel bullring built in 1898, on the Placa d'Espanya in Barcelona, into a multipurpose entertainment and cultural center for the developers SACRESA. Engineer Chris Wise and architect Laurie Abbott developed their initial ideas as they surveyed the site from the center of the bullring. They sketched a UFO shape hovering over the arched bullring, forming a domed roof, and even from the starting point of this conceptual doodle they emphasized preserving the essence of the historic building. They proposed maintaining the circular geometry of the original walls but also saw the potential to build vertically in order to house additional floor levels for recreation, increased visitor occupancy, and underground parking and shopping.

As the design work proceeded, the engineers began analyzing the existing, decaying structure and the city's underground infrastructure, including its metro tunnels. They also began devising methods of supporting the existing masonry shell during the construction process. The bullring sat ceremonially on a low hill, but the team opted to excavate five meters of ground from around the structure, leveling out the approach and linking it to the street. As Wise observed, "If you want to turn it into a contemporary place with a mix of culture, shops, restaurants, and bars, the last thing to do is to make it hard for people to get in. Rather than it being castle on a hill, we connected it back to the street, putting something magical in between."

The new upper level consists of a pavilion that is 350 meters in circumference and permits views out to the city and the mountains beyond, combined with a 90-meter-diameter performance venue in the roof. They designed a lamella structure for the dome, imperceptibly bolting together Glu-laminated timber beams of fir in a lozenge pattern that changes as the low-profile roof rises the 10 meters to the crown of the dome to an oculus 30 meters in diameter. Such a shallow roof would normally be hard to inhabit and more prone to buckling and deflections than a dome with a larger rise. This problem was resolved by the insertion of a "skirt" that acts as a tension ring, at the perimeter of the dome, intended to create additional usable space along the inside edges of the dome. The performance spaces are in the center, with cafés and services occupying the skirt area. The roof perimeter is divided into fifty-six sectors following that of the original facade points. The lightweight timber dome, which consists of numerous small pieces, will be built in a factory

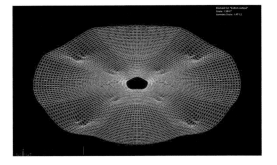
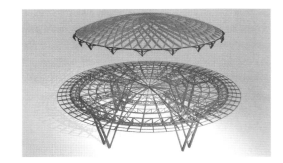

ABOVE: Initial sketches.

RIGHT: Computer analysis of the "dish" bending stress (above); steel-framed "dish" and timber grid-shell roof (below).

OPPOSITE: Axonometric rendering of new elements to be inserted into the historic arena.

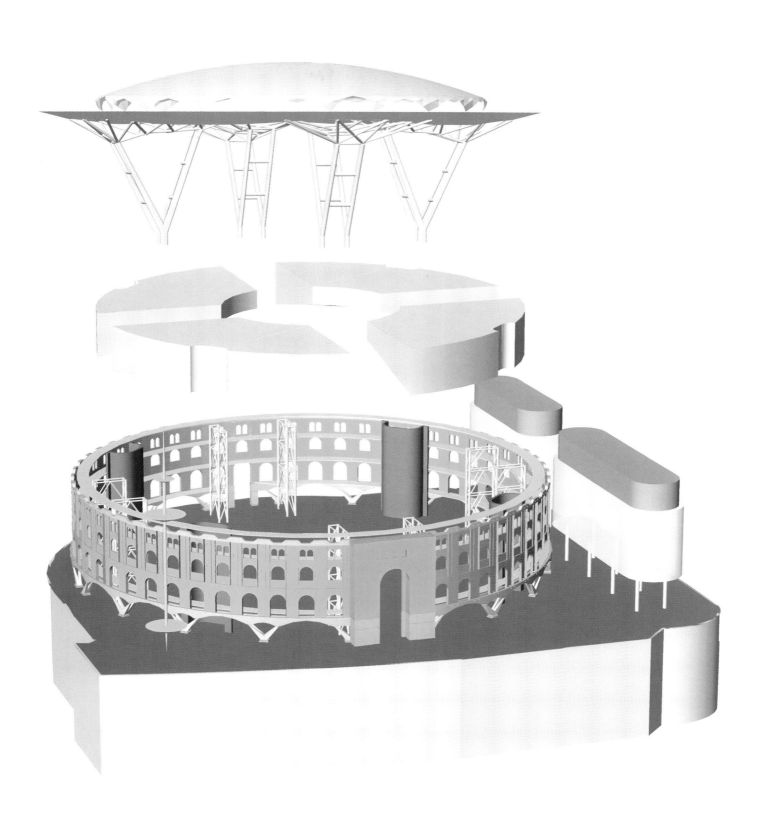

and then assembled on-site. The roof is supported by light cable trusses and is coated with white plastic that reflects sunlight and contains high performance thermo–solar energy collectors for hot water.

The dish-shaped floor of the performance space was shaped by means of a computer form-finding model and based on the concept of a soap bubble, on the principle that if you allow the structure to find its natural form, following the flux and flow lines, it will achieve maximum efficiency using the minimum amount of materials. Wise notes that he considers "the natural forces applied to buildings in terms of flow, not in terms of statics: they move." The dish is supported at four points with parallel, bifurcated columns, and through the form finding the structure strengthens itself where it needs to and then minimizes at the edges, defining the underside geometry of the bowl. A lightweight concrete floor is carried on metal decking, and as the two structures are linked together, they mutually support each other. The gap between the flooring and the underside of the structure became accidental spaces that accommodated the mechanical infrastructure.

The construction process was an engineering feat unto itself. Concrete ring beams clamped around the historic brick walls were supported by 5-meter-high steel vees. Similar in its arch geometry and forces to another Expedition Engineering project, the Northbank Footbridge, in Stockton, England, the force lines travel through and back down in a new arch, each 16 meters wide. Steel walkways circling inside the historic walls above double as additional stabilizing elements.

The overall concept for both the engineering and the architecture was that of a new structural system and new programming inserted into an existing structure. The effect is an explicit exposure of historic layers integrated with the new. Wise cites Scarpa's Castel Vecchio as a key influence in this approach.

TOP: Upper level support elements for roof structure; concrete arches supported by V-shape concrete piers; new structural supports under construction.

ABOVE: Seating in the original bullring.

OPPOSITE: Shell roof structure.

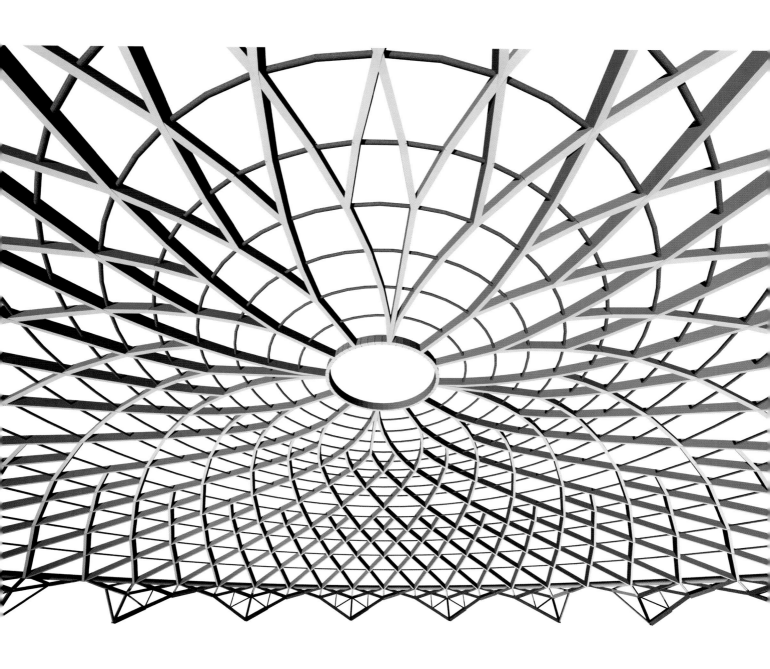

NORTH SHORE FOOTBRIDGE
STOCKTON-ON-TEES, UNITED KINGDOM

Expedition Engineering, with Spence Associates, won a 2004 Royal Institute of British Architects open competition for the design of a pedestrian bridge in Stockton, England, that will span the River Tees. Sponsored by English Partnerships, an organization dedicated to sustainable growth in England, the bridge, due to be completed in 2008, is part of a revitalization project for the Tees Valley and will connect the Urban Splash Teesdale Park Development on the northern river banks with the railroad station.

In the early stages, engineer Chris Wise and partner Ed McCann familiarized themselves with the site and then, on the return train to London, Wise played it back in his head, like a video, as he worked on the conceptual and structural design. And even in the first Stockton Bridge sketch, the rough concept of the engineering is implied. After an hour of drawing on facing pages of a large sketchbook, he developed the two-span arched structure.

The design evolved from the program and parameters of the site, so that the bridge spans the River Tees in two graceful arcs of different lengths, the longer span, a grand 120-meter arch on the northern end maintains clearance for the rowing course on the river, the shorter 60-meter arch continues across to the south side. As it narrows down at the embankment it extends into the town, continuing on an elevated path and tucking in between small buildings. A walkway continues along each shore underneath the bridge merging with the landscape. The arches spring from a single point and divide near their apex becoming deeper in section as they arc down to the supporting piers, which rear up from the river in the form of two 'X's. The legs follow the thrust line of the arches as they are resolved into the piers below the center line.

Each arch consists of two forked steel arches, essential to arrest any buckling and provide the necessary lateral support; the deck cables tie it all together. The fluid form of the upper portion supports and suspends the pedestrian footbed. A strong central support made of four piers grouped on sixteen piles, anchors the bridge in the river. From the center, the structure is tied together with cables and clamps the hanging deck at the ends. The cables are always in tension, like guitar strings.

Wise explains, "When I was at university I conducted a theoretical study on thrust lines inside arches; and you can draw where the forces are going. You can chase the forces of a vault in one perfect line from the apex to the ground, and as long as that line sits inside the geometry of the arch, it can stand up."

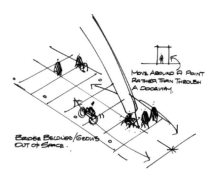

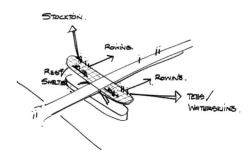

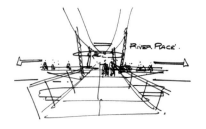

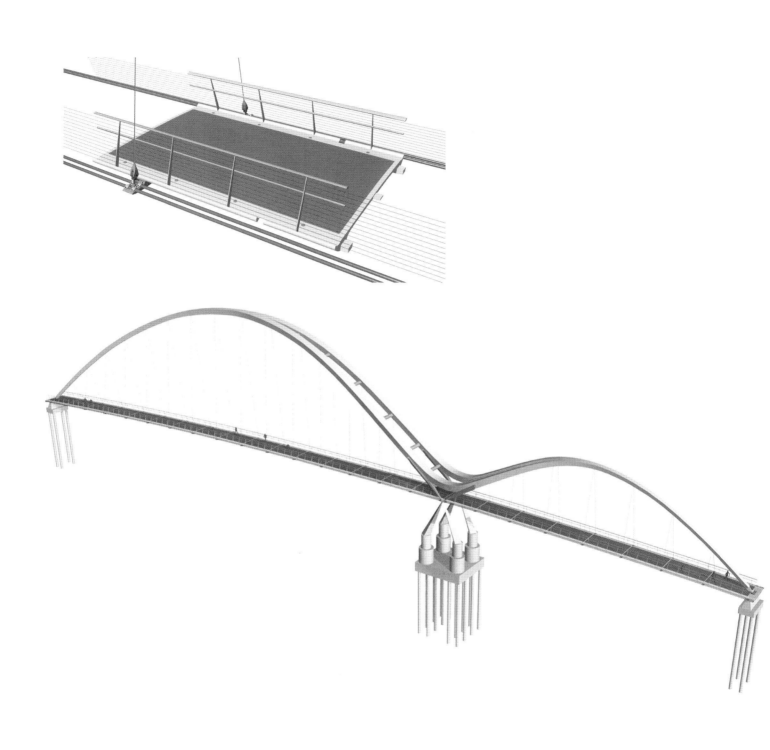

ABOVE: Deck details and structural supports in piers.

OPPOSITE: Sketches of the bridge site and context and rendering of the footbed.

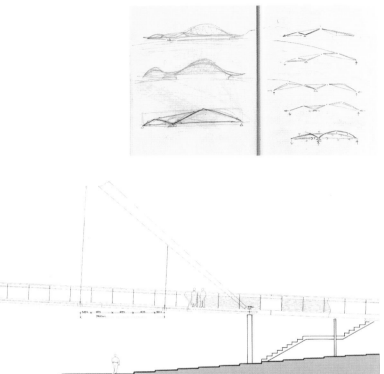

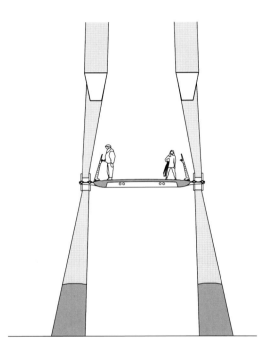

CLOCKWISE FROM LEFT: Elevation of site; sketches of bridge; elevation of the bridge and pedestrians; rendering.

OPPOSITE: Rendering of bridge.

The natural equilibrium contains this magic line and it has no thickness, which is the aspiration of the Stockton arch as it borrows from ancient traditions and forms an eventful urban passage.

"We are trying to get every element to do more than one job; the support cables and the ties that hold up the bridge are also a construction device. Individual elements are used to the minimum, which I learned from working with Norman Foster." They tested the force flow of the bridge using a chain model, and the bridge's slender arch system, which pushed the limits of engineering practice, was put through its paces in wind-tunnel tests. To counter wobbles induced by walking, it was determined that 5-ton tuned mass dampers should be added to maintain the bridge's comfort. The forces were the easiest of the issues to resolve, but the wind and dynamics, however, are more significant. And as Wise recognizes, "For an engineer, the proof is when it actually works in the physical world."

LESLIE E. ROBERTSON ASSOCIATES

Based in New York's Wall Street district, Leslie E. Robertson Associates (LERA) is best known for the ingenious engineering of the World Trade Center (Minoru Yamasaki, 1972–73) and other tall buildings. SawTeen See, William Faschan, Daniel Sesil, and Richard Zottola are the firm's partners with Leslie Robertson as the founding partner. The collaborative and collective partnership has allowed each of the engineers to develop innovative designs and fosters excellence in their work. The firm has invented numerous new elements, as points of innovation, that have changed the way buildings are built and designed, and Robertson himself is keenly interested in how ideas develop and then are shared. He says that "ideas are funny, they kind of float out there in the cosmos and they wait for someone to grab them and use them. Every idea has been used before; *you* just don't know it, to *you* it is brand new."

Robertson, who studied electrical engineering at the University of California, Berkeley, first focused on power distribution and towers but soon switched to engineer structures. He was only thirty-five when he began working on the World Trade Center with John Skilling of Worthington, Skilling, Helle and Jackson, and making the design happen presented Robertson with an exciting opportunity and challenge. The tallest buildings in the world at the time of their construction, the World Trade Center towers were also unique in terms of their cladding and structural concepts.

Before Robertson, the Chicago-based engineer Fazlur Kahn (1929–1982), who engineered many projects for Skidmore, Owings & Merrill, influenced new structural systems for skyscrapers for more expansive open floors by employing tube structures for buildings such as the John Hancock Center (Chicago, 1969) with its optimum column-diagonal truss tube that tapers, and the Sears Tower (Chicago, 1974), which uses a cluster of nine tubes to resist the wind.[1] Tubular structural principles are visible in bamboo, bones, and naturally formed hollow cores that support weight and wind. Khan achieved innovation for the tube stiffness by eliminating the structural grid and instead placed structural steel on the periphery, functioning as a beam with holes punched for windows.

For the World Trade Center, Robertson's firm developed numerous structural innovations such as free column space made possible by the tube structure, opening the floor from the service core to the outside wall; the most extensive prefabricated steel panel systems at the time (all of the steel components were fabricated off-site); sophisticated computerized Finite Element analyses using IBM punch cards; the invention of a viscoelastic damping system and a newly devised shaftwall fire-rated partition system.

In eliminating the more traditional load-bearing wall, the tube structure became load bearing, so that the main columns were placed in the building core along with the elevator shafts and the stairs, and the wind forces were taken in the column/spandrel construction of the outside wall.[2] The vertical mullions were knit together in a framework, and a girder grid system supported the floors.

The steel-plate panels of the perimeter wall were produced in Japan and shipped to New York by way of Seattle. The metal floor systems were fabricated in various locales and assembled into floor panels in New Jersey. This process was made feasible through a smooth workflow coordinated by means of a new, advanced IBM punch-card computer system and an in-house developed computer program. This marked the first time that structural designs were delivered to the contractors in a digital system.

The shaftwall system, a fire-rated partition system of gypsum wallboard and metal studs, was developed from Robertson's own building investigations to reduce the stack-action. By riding atop the elevator cars of various buildings, he saw that the standard partition system of gypsum block or brick leaked like a sieve; if the buildings moved, the steelwork moved, and as the structure above separated, a crack opened up at the top of the wall, making the partitions permeable and allowing too much air flow. Previously, Robertson had been interested in issues of static pressure in buildings, such as the chimney effect, and had written papers on how to address it.[3] With the shaftwall the top of the wall could slide in relation to the floor above and not open up, because it is built from the floor, making the shaft increasingly airtight and stronger than with other current methods of construction. Robertson presented these ideas to the Port Authority of New York and New Jersey, and the concept was developed with Tishman Research.

Because there was less masonry in the buildings, they would sway. Robertson completed studies with motion simulators based on the velocity and turbulence of the wind to ascertain responses to lateral sway. He discovered that the tolerance for oscillation in tall buildings was much lower than psychologists had predicted earlier. To counter this movement, he developed a viscoelastic damping system that was then manufactured to LERA's specifications. The dampers perpendicular to the main system could thus absorb the wind-induced energy of oscillation of the towers. For purposes of structural security, outrigger trusses were added below the roof for stiffening, and they also supported the 135-meter-tall television tower on one of the towers.

The Twin Towers had been designed to withstand the impact of a Boeing 707 that was not traveling at full speed and would be low on fuel close to landing. But the airplanes that crashed into the towers on September 11, 2001, were above their rated speed and their fuel tanks were at nearly full capacity—an unpredictable event of another dimension. Following the events of September 11, 2001, in which the World Trade Center was destroyed, Robertson spent many months working through the issues tied to the meaning of technology and society and to his structural work. He continues, however, to direct his energies into the vital engineering practice that has for many years been at the forefront of the profession. The destruction of any building, especially in a violent, tragic condition, is not an easy burden but unusual for an engineer is that it denudes the work in an abnormal way. Most situations for structural failure are rarely experienced in such a catastrophic way. Though the buildings no longer stand, the structural achievements endure. The disaster did, however, trigger new investigations into code requirements for fire and safety landings, sky lobbies, and exit systems.

During the design of the World Trade Center, Robertson was also working on the U.S. Steel Building in Pittsburgh (Harrison, Abramovitz, and Abbe, 1970). The building suffered at first from having too many structural elements. The 256-meter-tall triangular building with serrated corners was originally proposed to have columns every 4 meters that were attached back to the building at every floor. Instead, Robertson suggested eliminating two or three of the columns, taking the spacing out to 12 meters, and attaching the columns at every third floor. In making these alterations, LERA melded the structural and the architectural tolerances to make it buildable, which improved the structural efficiency, economy, and appearance. The firm used outrigger trusses in this project, to tie the perimeter columns down, and the space frame at the roof reduced potential swaying motion.

Robertson emphasizes strong collaborations with architects who he sees as essential to the inspiration of a project. Robertson notes, "When you begin to work with an architect who you don't know, you learn about their work and see the good and the bad. I always take the position that what an architect sees as good I see as good, and what he sees as not so good, so do I. But you have to be frank about structural potentials."

LERA has a close relationship with Pei Cobb Freed & Partners and has engineered such notable buildings as the Bank of China Tower (Hong Kong, 1990) for the firm. Robertson says his "initial idea" for the Bank of China Tower "was for a composite braced mega-structure, which hadn't been used before, and that was influenced by the wind loads. It was not done for structural gymnastics, but it improves the stability of the building and stabilizes it in case of a typhoon, and so that made good economic sense." As built, the tower has a 54-meter clear span rising up 370 meters. As a diagonal load-bearing structure, the exterior trusses of structural steel receive the horizontal forces, which are then absorbed by four composite steel-concrete columns at each corner. Refuge floors, required in Hong Kong high-rises that serve as intermediary floors or rest areas in case of fires, were used to fulfill fire-safety codes. LERA made planar trusses in steel with two different geometries but the X-shape trusses the team specified were at first not approved by the bank, because

ABOVE: World Trade Center, Minoru Yamasaki, New York, 1973.

OPPOSITE: U. S. Steel, Harrison Abramovitz and Abbe, under construction, Pittsburgh, Pennsylvania, 1970.

PAGE 122: Bank of China Tower, under construction, I. M. Pei, Hong Kong, 1990.

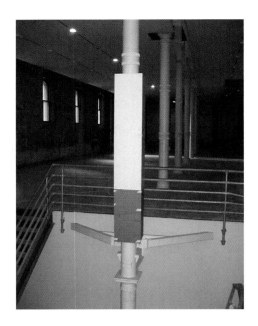

ABOVE: Prada showing column reconstruction, OMA with ARO, New York, 2000.
BELOW: Prada, OMA with ARO, New York, 2000.
OPPOSITE: Concept for Mile High Skyscraper, NBBJ, 2005.

of the negative connotation of the X. Instead, Pei called the structural frame "diamond" shapes, to signify wealth, and they were approved. The diagonals intersect at the corner of the facade, with the columns inside.

Often, Robertson says, he will have his own breakthrough in a technique—such as when he was able to make space frames into structural skin—but he sees working with other people as the spark for these ideas. For example, Robertson notes, "I. M. would say that the idea was mine, and I would say, 'It is not. It is our idea; it isn't my idea at all and it works.'" But regardless of where the ideas come from: "I do know that at night I wake up with realizations that I hadn't had before; I just wake up and say woo."

Dan Sesil, who has been with the firm since 1983, sees design as an evolutionary process, something that percolates over time. But, he says, "The evolution of an idea is contingent upon the interplay between creative players; quality design comes from the energies of a group. That is why designs evolve and you exchange six or eight ideas and they evolve into another six or eight ideas, and it becomes extraordinary." But he emphasizes, "there is not a great project without a great architect. Architects make a difference along with a collaborative process. But the most important thing is clarity."

For the Prada store in New York (OMA with ARO, 2000) Sesil had an opportunity to design a hybrid structure in an existing loft building. He used the original eight-foot-square brick-and-granite basement piers as the primary supports and as a vertical cantilever, supporting the new space from the pier while the building was totally occupied. In a simple, elegant move, LERA sheathed the column

in metal and integrated it into the foundation. LERA also designed the glass of the elevator cab and cantilevered the seats off the glass skin, developing new anchoring techniques in the process.

In considering how the role of the computer has evolved over the years, Robertson says he has seen the possibilities it offers for executing straightforward changes quickly. He also has observed the differences it has made in building performance and in the economies that result from computer analysis. But it is Sesil and his peers who recognize the computer's enormous power and potential as a design tool. "We spend a lot of energy thinking in 3-D and then spend time on the computer drawing it in 2-D just so that someone else can assemble it in 3-D again; the computer can allow us to skip that middle step." He is concerned, however, that the computer has been viewed as largely a geometry and form-shaping tool for engineers. "I can come up with ten ideas that we can analyze on the computer, but we don't need to make something just because we are able to; a project needs to be rooted in something more meaningful." Key to Sesil is the distinction between intuitive understanding and the power of analytical tools. "The computer is a great tool but there is nothing that can take the place of creativity."

Robertson continues provocative structural research in supertall buildings using formulas based on materials that are stronger and lighter in projects such as the NBBJ mile-high spiraling structure. But he notes that design of a supertall building cannot be based on engineering alone: "Each proposed structural concept must be weighed carefully against functional and aesthetic considerations." He urges structural engineers to make their hidden structures beautiful. "Even if no one ever sees what you have done again, you will be happy with it. Handsome structures are more efficient than ugly ones."[4]

Robertson considers the well-tested materials concrete and steel inefficient, and he awaits materials such as a carbon epoxy or fabrics that could be pressurized in space. For Robertson, the potential of material and structure are as infinite as the imagination. But he notes that "should structure alone be the deciding factor in establishing the form of such a building, the resulting edifice will almost surely fail to bring that essential sense of pride to those who create it and to those who make use of it."

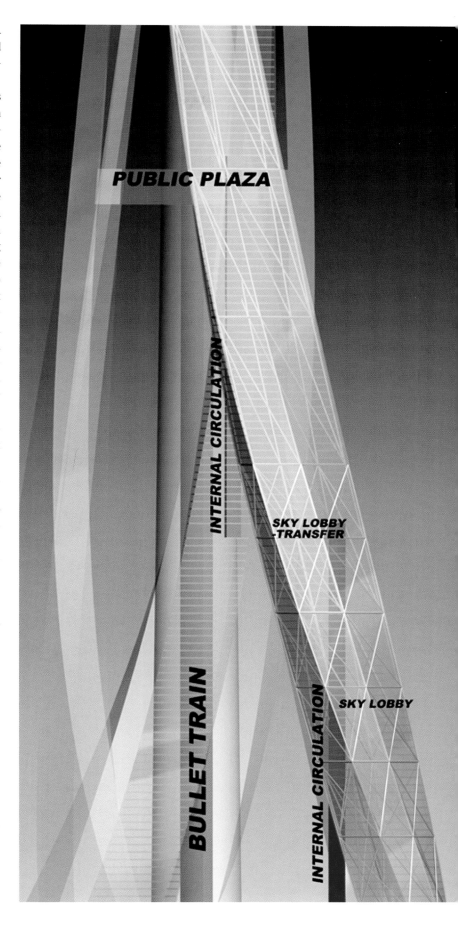

MIHO MUSEUM BRIDGE
SHIGA-RAKI, KYOTO, JAPAN

The Miho Museum Bridge, built to reach the 1,580-square-meter Miho Museum across a natural ravine within a nature preserve, transitions from a tunnel into a cantilevered bridge. I. M. Pei conceived the museum as a metaphor for the peach-blossom spring as described in a story common in Chinese literature, about a fisherman setting forth in the morning on a journey and coming upon Shangri-la. The message of the story is that the journey is as important as the destination.

The client, a spiritual group called Shinji Shumeikai, was restricted by the nature preserve in what could be built and what materials could be used. A height limit of 13 meters was imposed to minimize the visual impact of the building. Thus, most of the museum was constructed underground, emerging aboveground as hipped, temple-like glass roofs supported by steel space frames.[5]

For the Miho Bridge to achieve the shallow, slender profile—it is 120 meters long but only 2 meters deep—sought by Pei, Leslie Robertson developed a clean-lined, cantilevered structure. In essence, the tunnel transfers to steel beams and cantilevers out over the valley. Then there is a drop in section, and the bridge continues in one fluid gesture. The tunnel end comes out of the hillside, and its floor is supported in axial compression by a space frame. The steel chords of this hybrid cantilevered, cable-stayed, post-tensioned bridge

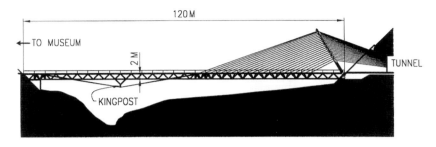

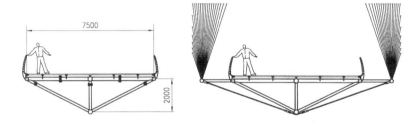

ABOVE: Elevation; plan at space frame; typical section and section at cables; structural diagrams of pedestrian deck and cable stays.

OPPOSITE: View from bridge to tunnel.

CLOCKWISE FROM LEFT: Fabrication of space frame; space frame connected with cable stays; view of bridge in landscape.

OPPOSITE: Aerial view of bridge from the tunnel entrance to the Miho Museum.

are 267 millimeters in diameter, and are anchored to the base of the tunnel. Post-tensioning worked to counteract the deflection and compensate for the anticipated live load and other forces. The deck was clamped against the abutments, and the bridge was lifted up by post-tensioning cables. This induced a desirable level of stress resulting in an elegant and dramatic bridge.

Designing a drainage system free of obtrusive pipes resulted in another structural innovation. Recalling a tennis court surface's porous ceramic infill, Robertson created a ceramic and stainless-steel deck that water could drain directly through. Convincing the client to use it was a challenge because they felt that the porous floor makes for a tentative walking surface. But the sample Robertson commissioned from a Japanese grating manufacturer turned out beautifully, and the steel decking is embedded with ceramic beads through which the run-off penetrates, with design innovation.

NEWSEUM/FREEDOM FORUM FOUNDATION WORLD HEADQUARTERS
WASHINGTON, D. C.

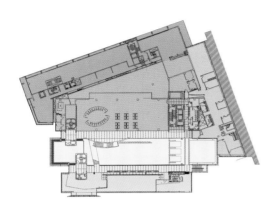

ABOVE: Ground floor plan showing the four spatial divisions called "bars."

BELOW: Second level plan and main facade window detail.

OPPOSITE: Pennsylvania Avenue facade.

The Newseum/Freedom Forum Foundation World Headquarters on Pennsylvania Avenue in Washington, D.C., designed by the Polshek Partnership, symbolizes the contract between the "Fourth Estate" and the government, as well as the bond between the free press and the public. The 550,000-square-foot mixed-use building includes not only a museum but a 22,000-square-foot conference center, 54,000 square feet of office space for the Freedom Forum Foundation, and 145,500 square feet of housing in nine stories that face Sixth Street. The museum is organized in layers moving back from Pennsylvania Avenue representing a three-dimensional newspaper. The first layer is the facade with a 60-foot-high carved stone tablet engraved with part of the First Amendment and a 60-by-80-foot "window on the world," a transparent glass wall through which exhibitions can been seen inside; the second is the public area, and the third the central core of the building housing exhibits and the theater.

Dan Sesil of LERA focused on integrating the main structural components in a way that would preserve an efficient and clear form for the building and its many parts and programs. Certain elements of the structure were especially innovative: the glass "window" wall, which opens up the five-story space; a 6,600-square-foot, column-free public atrium; and the stair/column along Pennsylvania Avenue.

The transparent 60-by-80-foot facade is emblematic of the value of openness and the responsibility of the press to society. It is also, literally, in the foreground of the structural exploration for the project. Starting with sketches and working with the Polshek Partnership and the curtain-wall consultant Robert Heintges, the engineers reversed the "tennis racket" approach to cable wall structures, in which the grid of cables sits close to the glass, and instead broke out the systems and isolated individual elements. Typically, in the gridded system, the cables would be anchored in the perimeter wall. For the Newseum, they departed from that practice and designed a one-directional

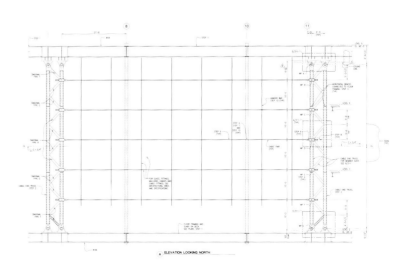

ABOVE: Glass corner under construction; vertical truss; stair tower under construction.

OPPOSITE: Stair tower under construction.

system to carry the wind load in the tensioned cables, which also simplified tuning the system. But the question of how to support the glass remained. They decided that by limiting the size of each glass panel—which are connected at their corners by nodes—only every other connection provides vertical support. The fixed vertical support offers stability, and the free horizontal support allows the necessary movement. Sesil describes the principle behind this: "This glass wall is organized around the idea that the glass panels work in pairs, and the two pieces have the gravity load taken up, and the wind is handled by the cables, so it is very crisp. For me this structural design is about taking proven technologies in an honest way that works with the space, not reinventing everything."

For the 90-foot high atrium space the engineers used a 252-foot-long truss to realize the architectural vision for horizontal flow and open space. "The top and bottom chords are the floor, framing at the sixth and seventh levels connected by two prefabricated web trusses creating a long box truss. The truss spans the building 148 feet from east to west. The system is a box supported on a reinforced concrete core tower to the west and on steel piers, 123 feet tall to the east." The tube was thus cantilevered 36 feet and 48 feet respectively at each end with the supports pulled in from the end that both reduced the materials needed and the bending moment. The floor acts as the bottom of the truss chord, the top chord is the ceiling, and the diagonal bracing is separated from the truss chord as a result of the tubular nature of the structure. The trusses thus enable the main atrium space to be column-free, with one floor suspended on the west end of the tube by the concrete core and two floors on the east. At the ends the expressed truss diagonals support steel hangers from which hang multiple levels of exhibition space at the ends of the truss. This allows for tall high spaces with natural light.

Another key structural element that influenced the building's design is the transparent glass corner at the east end of the first bar, where the stair acts as a bent column, leaving the corner clear of supports. In an illustration of the concept of load and reaction, each tread is supported by the next in moment-corrected steel box girders, and as the stair takes shape, it becomes a design element. Sesil emphasizes that he wanted to understand "the spectrum of where the efficient place in structure resides, because that can morph the design into a different shape. The deeper question is honesty, and how the piece can be most efficient and represent the process." He says that engineers should be able to plot an efficiency curve as it relates to the spacing between pieces and distribution of materials. The truss, the assemblage of plates, the thickness, and the spaces between resulted in an efficient, honest form that represent a philosophy of structure.

GUY NORDENSON AND ASSOCIATES

Guy Nordenson likens the work of an engineer to that of a jazz musician who plays a melody and picks up riffs and tunes from other players and transforms disparate elements into new, collaborative compositions. "Tunes get developed out of the momentary circumstances of the performance, which is true with any good collaboration: Engineering and architecture are not autonomous arts." Nordenson also compares the collaborative process to the poet Ezra Pound's idea of the vortex, a dynamic energy, or "a radiant node or cluster . . . from which, and through which, and into which, ideas are constantly flowing."[1]

After having studied comparative literature and civil engineering as an undergraduate at the Massachusetts Institute of Technology, he earned his master's degree in structural engineering and structural mechanics from the University of California, Berkeley. While still in school, Nordenson interned with sculptor Isamu Noguchi, and through that connection he worked with Buckminster Fuller.[2] He worked for Forell/Elsesser in San Francisco, in 1982 joined Weidlinger Associates in New York, and then in 1987 opened the New York office of Ove Arup & Partners. Nordenson was a director at Arup until he started his own firm, Guy Nordenson and Associates, in 1997. Since 1995 he has taught structural engineering to architecture and engineering students at Princeton University.

To Nordenson, engineering is the antithesis of autonomous art because design engineering is obliged to react to particular circumstances and situations. For him, projects do not necessarily begin with a blank piece of paper, and although they might entail creative formal thinking and brainstorming, artistic autonomy is more a myth of the modernist romantics such as Le Corbusier. Instead, for Nordenson projects evolve from a collective inquiry, as in science, and link back to Enlightenment and classical ideals and to the social fabric, rather than evolving from a singular idea.

ABOVE: Canopy at US Airways counter, LaGuardia Airport, Smith-Miller Hawkinson Architects, New York, 1993.

OPPOSITE: Kiasma Art Museum, Steven Holl Architects, Helsinki, Finland, 1998.

PAGE 136: Simmons Hall, MIT, Steven Holl Architects, Cambridge, Massachusetts, 2003.

One riff in design that Nordenson picks up is an attention to detail and materials that allows him to combine them in new ways under a variety of conditions. For the translucent canopy over the US Airways ticket counter at LaGuardia Airport (Smith-Miller + Hawkinson, 1993), Nordenson and the architects used a composite sandwich panel of epoxy and Nomex (honeycomb) reinforced with glass and carbon fibers, a material similar to that used for airplane and sailboat bodies. Prefabricated strips follow a stress pattern, resulting in a strong and thin shell. The composite shell then serves as the top chord of a truss of tempered glass, stainless steel cables and rods and a bottom chord of 3/4-by-6-inch carbon steel plate.

The physical and visual effects of material integration are also apparent in the Byzantine Fresco Chapel, in Houston, Texas (François de Menil, 1997), where an interior glass-and-steel structure echoing that of an original Cypriot chapel floats suspended from the church's ceiling, creating a space within a space. With an element as small as an interior stair in a New York City loft apartment (ARO, 1999), Nordenson demonstrates his novel use of material, integrating glass, steel, and aluminum with wood treads supported on one side by a glass panel 35 millimeters thick.

Often an engineer guides the architect's intention for a structure, and Nordenson has done just that in his consulting work. He has often collaborated with architect Steven Holl, on projects such as the Kiasma Art Museum, in Helsinki, Finland (1998), designing the truss structure that forms the thick torus-shaped roof and accommodates all the building services.

For the new La Chiesa di Dio Padre Misericordioso (Jubilee Church) on the outskirts of Rome (Richard Meier & Partners, with Ove Arup & Partners, 2003), Nordenson's firm helped the architect develop the three parallel shell-shaped walls, which foliate like plates in concentric spheres and which demonstrate how the engineering of the structure directly impacts on the sculpturing of interior space. The exterior curved precast-concrete blocks are stacked to heights from 18 to 28 meters, relating to the large block retaining walls and stone buildings of Rome. The technology is similar to engineer Pier Luigi Nervi's precast-concrete buildings, such as the Turin Exhibition Hall (1948). Nordenson's firm worked with local engineers Luigi Dell'Aquila and Antonio Michetti, disciples of Nervi's, and the project came together with an unusual synergy. The concrete mixture developed by the company Italcementi included photocatalytic particles that oxidize and become white when exposed to sunlight. The 8-ton, 75-centimeter-thick concrete blocks conceal the rods that connect the structure in a Freyssinet cable system, used for post-tensioning, anchored at the top and bottom.[3] The blocks are cantilevered from the ground upward and were erected with a mechanical gantry crane that sequentially rolled to each section. The slow motion block-by-block construction, like that of a medieval cathedral, became a community event.

On some projects, Nordenson is the design lead of the team, as with the 2002 competition for the Portland, Oregon, Tramway, for which Nordenson's firm invited New York architects ARO and the landscape designer Catherine Seavitt to participate, or the 2005 competition for the glass-canopied courtyard of the Patent Office building, in Washington, D.C. with Pei Cobb Freed & Partners. This reversal of roles and collaborative way of working, Nordenson notes, "sublimates individual satisfaction, because that is the way culture is built. The work is cultural and individual; it is the collective that makes a cultural contribution."[4] Nordenson, whose office is in lower Manhattan, had what one might call a revelation after the World Trade Center disaster in 2001. His practice had focused on individual buildings and structures—he does

not deny that he loves beautiful things—but he decided to pursue a deeper engagement with the civic side of engineering, civil engineering, as a social project. This work requires a different kind of problem solving and a military-like organization in order to bring the work to fruition, but Nordenson realized that he could make a contribution in this realm.

After September 11, 2001, the Nordenson office, drawing on its expertise in post-earthquake inspections, offered to conduct damage assessment around the World Trade Center site. Coordinated by the Structural Engineers Association of New York, the study included inspections, consulting on recovery and demolition plans, and detailed damage assessment of the surrounding buildings.[5] The firm also participated in a 2002 design study sponsored by the *New York Times Magazine* for new towers in lower Manhattan. Nordenson's firm later incorporated its own structural concept of a torqued tower into designs for the World Trade Center tower by Studio Daniel Libeskind and Skidmore, Owings & Merrill (SOM). In 2006 the firm also worked on the structural design for the bracing of the exposed slurry wall for the future World Trade Center Memorial Museum. The firm devised a liner wall structure that can be built on the outside of the slurry wall and tied with anchors, keeping the wall exposed but retaining the soil and water.

At the World Trade Center, Nordenson had the opportunity to explore the interaction between architecture, infrastructure, social structures, and democracy and the connection between the organization of spaces, human rights, and freedoms. The crisis conditions forced quick resolutions

ABOVE: Jubilee Church, Richard Meier & Partners, Rome, Italy, 2003. Concrete block mechanized construction system (left); erection of curved wall (right).

OPPOSITE: Sketch for World Trade Center Tower 1, New York, 2003.

and it became, for just a short time, a communal endeavor: "It was like the Paris Commune with a sense of collective effort, a new beginning of democracy."

As the controversy increased over the design of the Freedom Tower, David Childs of SOM asked Nordenson to be involved in its engineering, and the firm completed a schematic design for the tower in May 2003. Nordenson continued working on the project until December, when SOM with developer Larry Silverstein presented the final design. According to Nordenson, he based his design for a torqued tower on the manipulation of the parallelogram site geometry to create a synthesis of complex sculpture and structure. Creating a partial twist, the structural geometry of a diagrid framing system (which consists of intersecting diagonal network) could rise up to an open cable structure at the top floors. He notes that the twisted pylons and diagrid bracing references Fuller's tensegrity structures. Nordenson, who cocurated the exhibition "Tall Buildings" at the Museum of Modern Art with Architecture and Design curator Terence Riley in the fall of 2004, emphasizes that some tall buildings are obelisks and others, such as the Hong Kong Bank of China and the Sears Tower, are asymmetrical and figural, with a front and back. "This torqued form is based on that compositional effect. It was a happy consequence of the site geometry that you could get this form." The 200-foot parallelogram also mimics the shifting street grid of the city. With two opposing, or "held," corners as a vertical, and two sloping corners, it tapered as it rose to its 2000-foot height.

Incorporating the required antennae into a twin mast of two cylindrical concrete cores, stayed by a perimeter array of cables, created an overall effect similar to the Statue of Liberty, with its central core, outrigger and bronze wrapping. The torque would turn the top of the tower to face the prevailing northwest winds, allowing the effective use of wind turbines; and the tapering would follow the central elevator core with enough room to hold stair and office depths of 45 feet. The structural design consisted of a central core with a steel diagrid bracing system to 1,000 feet, and a cable net rising to 1,700 feet, where a truss was to provide support for the 300-foot antennae. The diagrid eliminated the need for vertical members. The floor-to-floor height is 13 feet, 4 inches.

For the World Trade Tower 1 site proposal, the twisted form became the volume, and Nordenson saw it as "a manifestation of a form of opposition to the reactionary, sentimental, and expressionist kitsch of the master plan proposal."

As becomes evident in Nordenson's work, culture, societal values, and politics figure just as prominently in conceptions of a built structure as do space, form, and light. And engineers and architects who make critical inquiry part of their process enhance the culture in general and achieve invention in their projects, through engineering culture.

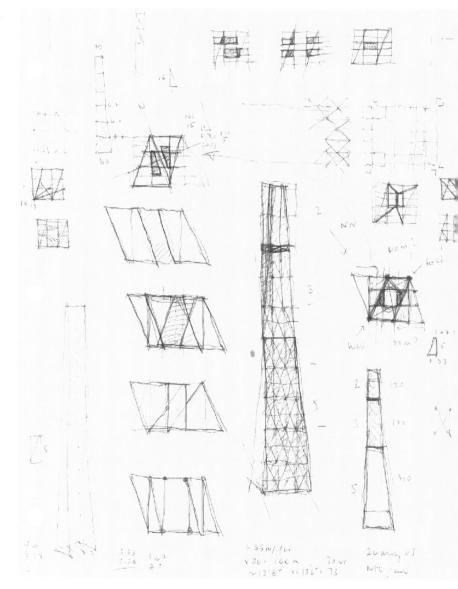

SIMMONS HALL, MIT CAMPUS
CAMBRIDGE, MASSACHUSETTS

For a new 350-student dormitory building along Vassar Street at the edge of MIT's new, expansive campus development, Steven Holl explored a number of formal compositions: a podium with towers; a "folded street" with ramps around the cores, bracketed trusses, and attached tubes; and a building spongelike both physically and metaphorically. This last design was selected for implementation, and the Boston-based firm Simpson, Gumpertz & Heger, which had worked with Buckminster Fuller on the 1967 Expo Pavilion in Montreal, was engaged as consulting engineer, with Guy Nordenson and Associates as lead engineer. The project was a case study in an empirical, open approach to problem solving and facilitated a good working relationship among the contractors, fabricators, and designers. Whether or not the sponge metaphor continues to ring true, the building and its structural system constituted a unique and dynamic endeavor. The overall building mass was one long, narrow, but massive block 105 feet and 10 stories tall, 385 feet long, and 53 feet deep, with corners recessed by means of cantilevers and open spaces and atriums carved out of the gridded structure in odd shapes, making the building appear porous, though not penetrable from front to back.

The three-dimensional gridded matrix could be described as a habitable Sol LeWitt sculpture, its spatial array formed by horizontal and vertical members originally designed to be the same size, though the bottom half of the building changed in depth. Nordenson's office suggested using precast-concrete panels of 10 by 20 feet to include the window openings, fitting the panels together like Lego blocks from the top of one panel to the bottom of the next. These were joined by the cast-in-place concrete floor slabs at each floor level. The rigid but flexible system meant that the design could begin with a standardized interior bedroom layout, and voids of various shapes could be blocked out to accommodate larger common areas and study rooms through a customized prefabricated concrete system.

Prefabricated concrete was the preferred material to be exposed inside, both for precision and surface quality, and adaptability. It was clad with aluminum on the exterior. Different versions of precast-concrete panels were developed, and the panels came together, one panel per dorm room, with window openings set in a square of three windows over three windows per room. The rebar extending through each end of the panel was threaded through corresponding holes in the panel below and cemented with grout. A panel was connected every thirty-five minutes.

In the design of the living spaces, Holl decided where windows were needed and the team developed a series of rules that could be correlated with the structural need for occasional solid panels. Structural requirements were computed in an iterative process in which the grid was adapted to the variation of stresses, and where the stresses were found

LEFT: The complete set of precast concrete panels for the building perimeter.

OPPOSITE: Grand lobby staircase.

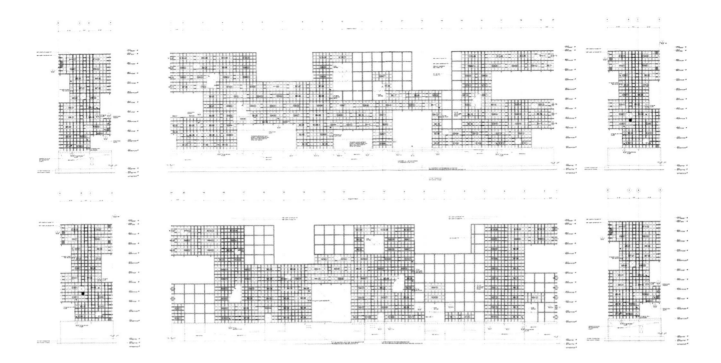

ABOVE: Color coding based on steel reinforcing bar sizes in precast panel horizontals and verticals, on the south (top) and north (bottom) elevations.

OPPOSITE: Detail of concrete panel construction system; concrete panel going into place; north facade.

to be greater, window openings were filled in with solid panels. Five large-scale openings in each of the ten dormitory houses break down the scale of the building. The grid spans these voids carved out of the interior and at the corners as well as the large corner cantilevers and other cutouts, functioning as a Vierendeel truss. Each time a window block was filled, the structural analysis program was run to recalculate the structural requirements in the region and the support moved to the optimal location for stabilization, but the sequence was different each time. As Nordenson said, "We could find one pattern and then start over. It was a nonlinear methodology. We liked that indeterminacy of the outcome based on how you activate the information." Nordenson enjoyed the challenge of making real the metaphor of porosity in Simmons Hall, because it appealed to his own interests in structural indeterminacy and allowed him to "do something whimsical" while "also meaningfully making visible a systematic way of dealing with variation." The structural expression results in an exoskeleton building in which the panels are a structural skin, in a new way of integrating structure and form.

The project's environmental aspects were engineered with Mahadev Raman of Arup's New York office. Natural ventilation was achieved through the large number of operable windows, including the nine in each dorm room. Insulation was added between the concrete panels and the exterior anodized-aluminum cladding. The interior concrete walls serve as a thermal mass, and the windows were set deep into the wall, blocking the direct sun.

Amy Stein, the site engineer, drew a map of the steel reinforcing bar sizes in the precast elements, which she used during inspections and color-coded according to the various diameters of bars. Nordenson suggested this color-coding as the basis for the color scheme for the building exterior that Holl had envisioned. Each color was keyed to the different sizes in the structural-reinforcement system. Blues signify the use of #5 diameter reinforcing bars, for example, yellows #7, and reds #9, making explicit the significance of structure in the design. As Steven Holl notes, "In most buildings, structure is about 25 percent of the cost and 30 percent of the material. So if you don't bring structure to bear on whatever it is you are doing as an architect, then I think that somewhere it is not working. It is not reconciled. Le Corbusier always incorporated the structure, as did Kahn and Mies. I am interested in maintaining an integral relation of the idea through the structure, the space and the experience."[6]

MUSEUM OF MODERN ART, GARDEN WALL AND BUILDING STRUCTURE
NEW YORK

Since the Museum of Modern Art had already undergone numerous accretions over the years, one goal of the 2005 expansion and renovation designed by architect Yoshio Taniguchi was to create a more unified complex. Other aims were to reconnect the building to the city with a through-passage between Fifty-third and Fifty-fourth Streets and to retain the original Edward Durell Stone and Philip Goodwin building (1939) on Fifty-third Street.

As completed with two new buildings, new galleries surround a central atrium, with the second floor featuring 21-foot-high ceilings and expansive galleries for large-scale artwork. A monumental staircase connects the fourth and fifth floors.

Guy Nordenson and Associates was hired as consulting engineer during the project's schematic and design development phases and also worked on the construction documents for the gallery stairways, curtain walls, and interior gallery structure. The main focus of the firm's work was the problem of creating structural frame geometry for the buliding and transparent and unobstructed glass walls for the three facades of the building that enclose the Rockefeller Sculpture Garden. The aspiration was to make the garden wall structure almost invisible. In keeping with the clean geometry of the building, the curtain wall is divided into a vertical rectangular grid 10 bays wide and 5 bays high. The

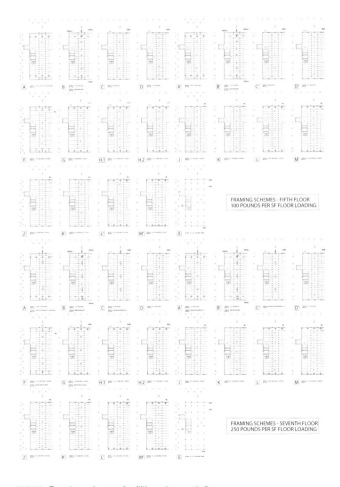

ABOVE: Framing schemes for fifth and seventh floors.

OPPOSITE: Rockefeller Sculpture Garden courtyard, west facade.

vertical solid steel mullions are 2.5 by 7 inches and rise 64 feet, supported at the third floor balcony and pinned at the top with slotted connections to resist horizontal loads. Mortise and tenon mechanical connections are hidden within the horizontal mullions, limiting the number of welds. Brackets are set flush, fastened with high-strength bolts that are concealed in keeping with the overall minimalist aesthetic.

To avoid transferring lateral racking movement between the existing tower and the curtain wall, spring connections are employed at the edge mullions on the third and seventh floors in an independent bracing system. A small fin cantilevers off the edge mullion and supports a ball bearing connection for the concrete tower structure. These small maneuvers made a thin veil-like curtain wall become the silent structure that the architects had desired.[8]

The western edge of the building is a braced frame verticle core against the building edge with a second structural braced line opposite on the east. The two vertical trusses are tied at the eighth, mechanical floor, and the top, sixteenth floor with belt truss structures for lateral strength and stiffness. From the mid-level belt trusses several columns are hung, opening up the contemporary gallery space on the second floor.

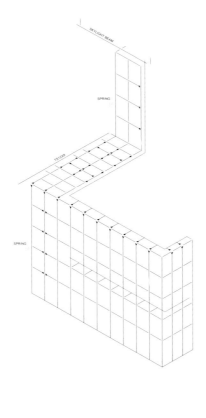

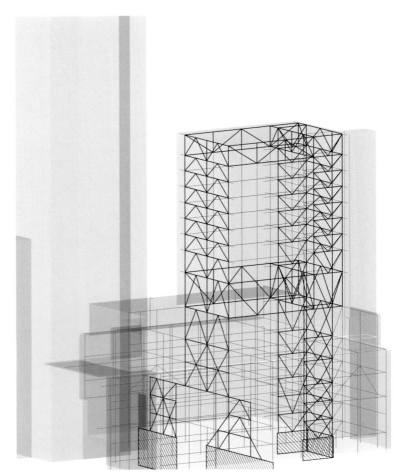

ABOVE: Diagram of glass facade details (left); axonometric with gallery truss at sixth and sixteenth floors (right).

OPPOSITE: Construction of facade structure; Lobby.

GUY NORDENSON AND ASSOCIATES

MINISTRUCTURE NO. 16
JINHUA CITY, CHINA

In Jinhua City, China, near Shanghai, Guy Nordenson and Associates designed the structural system for Michael Maltzan's 120-square-meter Ministructure No. 16. Situated in a park among a series of pavilions designed by an international group of architects, the ministructure provides a place for visitors to read, eat, and think. The concept is based on an important confluence between the book and architecture in Chinese history when in the fourth century B.C.E., Confucian writings were found after having been concealed in a wall during the third century B.C.E., when the emperor had ordered all Confucian writings burned.

Maltzan had wanted the pavilion to be made of poured concrete, a material he had not been able to experiment with in seismic zones in California. Guy Nordenson's first design had a perforated concrete wall as the structure; local engineers, however, determined that the building had to be made of steel because of the high water table and the need for additional support.

As soon as the building went to steel, Maltzan and Nordenson started discussing how to make the steel more visible and part of the texture of the architecture. It became complex because Nordenson had to design a structure around the design, a prominent feature of which was a perforated wall, and

ABOVE: Model showing steel elements.
OPPOSITE: Structural plan of the main floor.

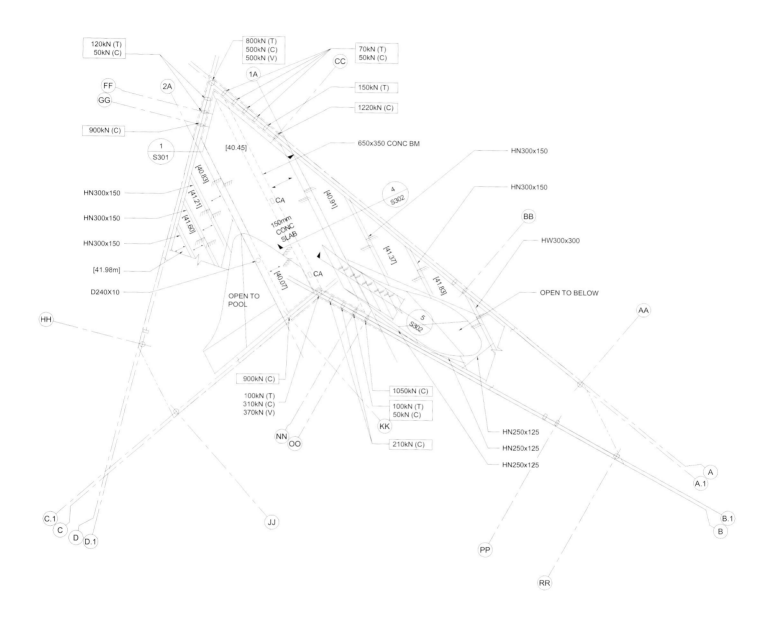

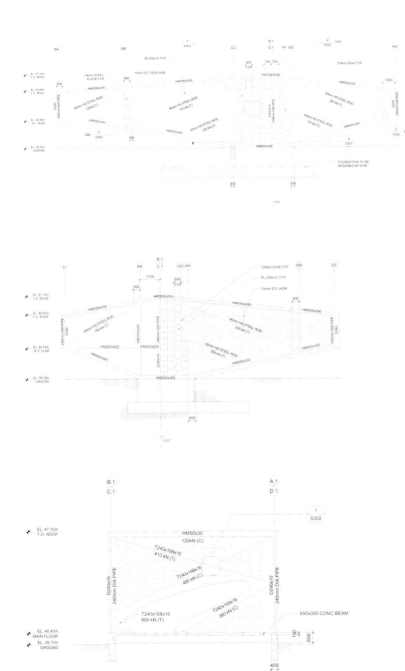

develop a hybrid Vierendeel structure with localized ladder trusses and diagonal pencil members. Together Maltzan, Nordenson, and the project engineer Erik Nelson rethought the skin, designing instead a double layer of slotted perforated metal panels with a larger rectangular opening. The structural steel is visible between the planes of the wall.

Maltzan says, "As opposed to the architecture leading decisions about the structure, it inverted halfway through the process and the structure led to the development of the project. Not because of the form, because that was set, but the skin itself. Then the new double skin started to make moiré patterns and it became animated through the structural design."

The form pulls its central wall outward into two unequal, cantilevered arms, each concealing a public space. The shorter of the wings contains a bookstore and café, organized into a series of terraces that rise to frame a view of the park to the west. The steps form a recessed space for books and seating, and an open reading porch, the underside of which forms a passage through which visitors enter. Although the bookstore and reading porch are separated by the center wall, the looping lines of three sinuous elements provide a connection between the two distinct spaces: the entrance aperture on the underside of the structure, its corresponding projected void in the center wall, and the curved plan of the mezzanine in the bookstore. The bent, tapered form seems to expand and contract, its perforated walls and openings creating different spaces at each level of vision and visibility.

LEFT: Pavilion under construction and completed (below).

OPPOSITE: Structural framing of north facade, south facade, and center wall.

GUY NORDENSON AND ASSOCIATES | 153

RFR

Peter Rice (1935–1992) established his Paris-based firm RFR in 1982 as a "laboratory and interface between architecture and engineering," after having worked for Ove Arup & Partners in London for twenty years on projects such as the Sydney Opera House (Jørn Utzon, 1957–73); the Pompidou Center (Piano & Rogers, 1971); and the Louvre's glass pyramids (I. M. Pei, 1985 and 1991). Rice, a visionary and humanist, was influenced by Ove Arup's "total architecture," the integration of design and structure in a holistic process. In distinguishing between the engineer and the architect, Rice said, "The architect's response is primarily creative, whereas the engineer's is essentially inventive."[1] RFR's workshops within the firm have exemplified this inventiveness, often beginning with a specific innovative detail that contributes to a total design.

After Rice's death in 1992, the RFR partnership continued to thrive and grow, as did the scope of projects. The firm is now directed by Kieran Rice (son of Peter Rice) and nine other directors including Henry Bardsley, Jean-François Blassel, and Bernard Vaudeville.[2] An international staff of eighty engineers and engineer-architects collaborate on projects that range in scope from conceptual design to construction-site supervision and that increasingly incorporates environmentally sustainable technologies. Multidisciplinary project teams are headed by a project director, but the firm's overall approach is one of teamwork, with collective reviews of projects and group brainstorming at strategic points.

At RFR new technologies—including new computer technologies and programs—are a means to innovation as dictated by the specific needs of each design, but not the dominant force. "The engineer is located between calculation on the one hand and experience on the other," says Blassel, "between audacity and rigor, but also between the risk of technological folly and the yoke of total control." The firm does not repeat solutions to technical problems, which makes the process of project

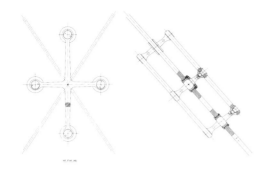

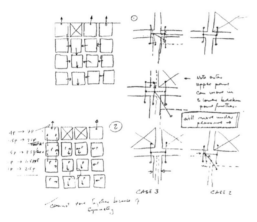

ABOVE: Sketch of star-shaped glass fastener for the Inverted Pyramid (top); Rotule system for the City of Sciences and Industry, La Villette, Adrien Fainsilber, Paris, 1981 (second from top); sketch for facades (second from bottom); H-Star glass fastener (bottom); completed facade (right).

OPPOSITE: Charles de Gaulle Airport, Terminal 2F, Paul Andreu, Paris, 1998.

PAGE 154: Inverted Pyramid, Musée du Louvre, I. M. Pei, Paris, 1993.

selection part of their ideology and a means of furthering the firm's growth.

RFR became known in the 1980s for innovative structural systems that revealed themselves in subtle ways and allowed observers to follow the cables, masts, supports, and tension rods to their points of contact and connection and visually comprehend the system. Peter Rice explored these strategies early on in his 1971 work on the design of the Pompidou Center as part of Arup's Structures 3 team with Ted Happold. Structure was clearly expressed, too, in the glass-and-steel facades Rice designed for the City of Sciences and Industry, La Villette (Adrien Fainsilber, 1981).[3] Rice also had a synergy working with Renzo Piano with whom with he experimented with materials such as ferrocement and ductile iron in projects such as the Menil Collection in Houston, Texas (1987).

More recently, at the high-speed (TGV) and commuter-rail (RER) station at Charles de Gaulle Airport in Paris (Aéroports de Paris, Paul Andreu, and SNCF architect Jean-Marie Duthilleul, 1991)—a 22,000-square-meter facility that floats above the underground train tracks—RFR combined glass and steel to create a novel and visible architectural effect for the roof, which opens up at its periphery. The design concept was to keep the glass smooth and lightweight, so RFR created a hierarchical composition in superimposed layers. Each is differentiated from the previous one not only by the progressive reduction of the slope, but also by the distinct geometry that creates the form while maintaining the architectural integrity of the project. Two cast-steel elements articulate the cluster of the tubular pylons, and a crescent beam is supported by the double central support. It is held in place primarily by the pretension of the vertical ties at each of its extremities, so the upper membrane is never in compression. Steel rods support the roof, and crescent beams follow the curve of the roof. The separation between the glass sheet and the crescent beams permits the elimination of expansion joints. The roof glass is fritted with a white mineral ceramic coating to soften the sunlight and

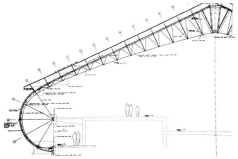

reflect the night sky, creating a luminous, atmospheric effect.

Structure is again revealed as a design element in the 8,500-square-meter Terminal 2F at the Charles de Gaulle Airport (Paul Andreu, 1998), which accommodates fourteen aircraft in a podlike, peninsular glass building. The 200-meter-long transparent peninsula on a solid base resembles an upturned boat hull because of its continuity—in this project expansion joints were not needed—and is fixed to the existing concrete structure at a single point along the ridge. The movement is controlled by articulated arm connections, which creates a sense of floating appropriate to an airfield. The continuous 50-by-10-meter transverse ribs have slender tubular-steel arcs stabilized by a fan-shaped array of tension rods at each side and become trusses that meet the apex at the longitudinal ridge beam. A perforated, folded stainless-steel plate above the glass forms a screen to limit solar gain. RFR developed a nonlinear computer model of the peninsula and refined and optimized the sizes of the elements after detailed analyses. The calculation models they developed for the complex geometry were the firm's most elaborate to date.

The construction of the west facade of the Cathedral Notre Dame de la Treille, in Lille (1860), was Peter Rice's last project, completed in 1997 after his death. The west facade and rose window, as with many cathedrals, was unfinished. Artist Ladislas Kijno designed the rose window and the doors of the nave and, together with architect Pierre-Louis Carlier, created a contemporary response to Gothic ideals that evoked the spirituality of the church. RFR developed a blue limestone Gothic arch from which they suspended a thin marble facade as a veil-like wall that both respected the existing church and was introduced as a modern technical innovation. Natural light penetrates the thin marble, transforming the solid material. An exterior steel framework of tension cables and rods supports the stone facade, keeping the stone in compression. A cross-shaped piece holds the marble panels together at the nodes.

In his book *An Engineer Imagines*, one of the few written by an engineer about his or her own work, Rice discussed the role of the engineer as making "real the presence of materials in use in the building, so that people warm to them, want to touch them, feel a sense of the material itself and of the people who made and designed it. To do this we have to avoid the worst excesses of the industrial hegemony. To maintain the feeling that it was the designer, and not industry and its available options that decided, is one essential ingredient of seeking a tactile, *traces de la main* solution." Rice also wrote that "the most powerful way that an engineer can contribute to the work of architects is by exploring the nature of the materials and using that knowledge to produce a special quality in the way the materials are used." He was conscious of how "properties of materials have dominated the way the design choices were made"[4] and the innovation needed to move past what is considered the normal or standardized use of a material.

Rice understood a material's ability to create ambiance. For example, he was concerned with the scale of fabric structures and with the way a fabric's cut and piecing determined its appearance and revealed the nature of the surface to give it a physical context with depth and body in space, as in Nuage de la Grande Arche (The Cloud at the Grand Arch) in La Défense, Paris (Otto von Spreckelsen and Andreu, 1983). And with glass, Rice achieved a lightness that created a sense of nonmateriality as the structure was made apparent through the transparent skin.

This sensibility still informs RFR's work, and combinations of materials and technologies—steel and glass, fabric and steel, stone and steel—remain key. "But what distinguishes engineering today from the past," says Jean-François Blassel, "is both the profusion of diverse materials and the erosion of the very concept of material, for indeed, the distinction between materials, structure, and envelope is becoming blurred. New composite materials have microstructures that allow them to

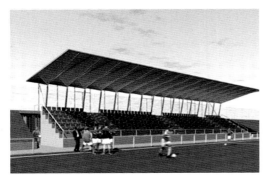

ABOVE: Nanterre Stadium, Bathélémy & Grino Architects, Nanterre, France, 2003. Grandstand and roof detail; Georges Restaurant, Pompidou Center, Jakob + Macfarlane, Paris, 2000.

OPPOSITE: Chambre des Députés, Luxembourg, 2006. Studies for footbridge and views of the atrium connection.

form a continuum with the large-scale, or macrostructure. What is more, they integrate complex thermal and acoustic functions, which formerly were the attributes solely of insulation materials. Their coordinated functioning leads to a diminished role for structure, which comes thus to seem no longer an entity in itself, and even less, the construction's skeleton."

The firm also focuses on computer fabrication processes as a way to precisely engineer a project. Henry Bardsley often focuses on what he sees as the dual nature of engineering projects, their intangibility and tangibility, where the fabrication process is part of the design and modeling from computer to mold to building element. He has noted, "Complex geometries and the tools that you have make the building, so that the building is a result of the tool and what that can make." At the small stadium in Nanterre, for which the firm designed a wood and polycarbonate roof, the computer numerical control (CNC) tool changed all of the chamfers so that they could tilt in different directions by using a one-dimensional milling machine. But on the other hand Bardsley notes that there is a contradiction today: the computer does the calculating, but the production is in the hands of a blacksmith, who makes the negative shape that then is hand tooled to make the positive element for the structural element or piece to fit. Handwork and craftsmanship still come into play.

In small-scale experimental projects, RFR has explored the need to protect natural resources and investigated the issue of economy of means by using alternative materials to create light, flexible, and temporary structures. One of these projects, the Birdhouse—so named because it was inspired by the flight of a bird—is a suspended polyhedral cube shelter that the firm both designed and engineered. The materials used—reeds, twigs, vines, and bamboo—are easily renewable. The Birdhouse generated its own requirements for technique and craftsmanship, as well as its own aesthetic expression and concern for renewable materials.

A small insertion into the Pompidou Center in Paris for the rooftop Georges Restaurant (Jakob + MacFarlane, 2000) demonstrated experimentation with form finding at a small scale, combining computer-modeling methods with aluminum-welding techniques borrowed from shipbuilding. Undulating metal rooms, or bubbles, were designed with a nonlinear geometry. Their metal pieces were milled to optimize the number of cuts and to more precisely position the structural elements. The aluminum hull is structural, and is covered with a sculptural skin. The joints were independent of the overall fabrication and are decorative and informative, exposing the modes of construction. The floor is a 1,400-square-meter smooth surface that was raised from the ground and wraps the whole plan as an undersurface, turning the project's volumes into fluid forms. Brightly colored welded plastic tiles form a continuous film along the interior surface. The bubbles stand on independent supports so as not to disturb the existing building's structure.

In a 2006 project that consisted of small, incremental insertions, RFR designed glass bridges and atriums connecting the Luxembourg Parliament Buildings. One bridge consists of a simple glass tube and has a carpetlike effect. Two steel beams handle the dead weight, and the glass box carries the wind loads with as little steel as possible. The flanking curtain walls form a double-height space and connect the landing and antechambers of the existing buildings. A glass roof over the courtyards has a thin skin of photovoltaic cells that doubles as a shading device. The vertical operable blades of the louvers are jewel-like, with every structural member custom crafted in metal, steel, and glass. The roof snakes through all the interstitial spaces, linking the entire complex physically, and also texturally, its glass playing off the slate of the roofs.

In contrast to the display of heroic engineering in the landscape of the industrial revolution, RFR's infrastructural projects achieve a subtle integration with the landscape. The design of a double viaduct for the TGV at Avignon with landscape architect Michel Desvigne and engineer Michel Virlogeux (who engineered the Millau Viaduct with Norman Foster, 2005) opens up the regional plan while having only a light impact on the landscape. The Paris–Marseilles and Marseilles–Montpellier lines converge, almost touching, on a viaduct of nineteen prefabricated concrete piers spanning the Rhone River valley. Piers are up to 100 meters apart, and the length of each of the two viaducts is 1,500 meters. Henry Bardsley acknowledges that designing a traditional precast-concrete corbel bridge—the first for trains—with a steel form in a particularly rigorous geometry was a sobering experience. The knowledge gained from meeting challenges posed by topography was put to good use in later canal and viaduct projects, as well as in 2006 proposals for desert road and bridge networks in Abu Dhabi.

Working with architects with different aesthetic preferences, most recently on projects with Frank Gehry and Kohn Pedersen Fox Architects, RFR's consistent approach has been to seek specific solutions for each situation rather than to develop a stylistic vocabulary. But the firm's work is not neutral: form is resolved in a process in which elements contribute to the final effect.[5] RFR remains committed to the concept of the architect-engineer: balancing intuition with rigorous mathematical equations, aesthetic concerns with structural rationality, pushing themselves to discover the unexplored potential of materials and structure, always tactile and comprehensible.

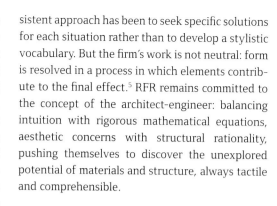

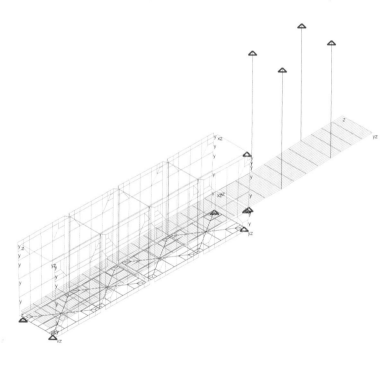

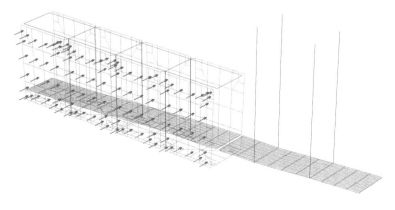

SIMONE DE BEAUVOIR FOOTBRIDGE
PARIS

The Bercy-Tolbiac Footbridge, named the Simone de Beauvoir Footbridge, over the Seine in Paris, is a slender 270-meter-long pedestrian bridge, the commission for which RFR and Austrian architect Dietmar Feichtinger won in a 1998 competition. The bridge, completed in July 2006, leads from the Bibliothèque Nationale de France on the left bank to the park at Bercy on the right, uniting two redeveloped urban areas. RFR engineers Bernard Vaudeville and Valérie Boniface worked closely with Feichtinger on the hybrid structural design, which enables the bridge to cross the river without an intermediary pylon, in one 190-meter free span that lies low on the landscape.

Two opposing synergetic elements, a shooting arch in compression and a suspended catenary arc in tension, equilibrate each other and create what the engineers call a "fish beam" or "lens," making the bridge both robust and light, visually and physically. The steel curvatures connect to three parallel aprons that follow the structural forces—one lifts up at the center with the arc, and the lateral ones follow the camber of the lateral catenary ties—to minimize the bending moment in the arches. The three different parts—consoles, lens shape, and vertical columns—are realized as a semi-Vierendeel truss or ladder-like steel structure, but slender and similar to a Gerber system, having two consoles and one isostatic beam. The interweaving of the structure creates double walkways for the deck of the bridge that correspond to the truss's lines of axial forces, formed by the arch and the catenary. The arches and their supports coalesce into a boomerang-shaped ensemble with the bracketed arm and leg (horizontal and vertical supports) and then are pulled back at an "elbow" for stiffness and anchored to the embankment. This allows the bridge to be isostatic, in equilibrium, from the principal structure and places the loads between the foundation and the arch. The bridge achieves an open quality as it ribbons through the cantilevered support system and anchors in the riverbanks.

The construction phase took more time than expected because additional movement analysis, using new techniques, needed to be completed. The extra measures were taken in response to detection of unsafe vibration in the Millennium Bridge (Norman Foster and Arup, 2000), a pedestrian bridge over the Thames in London, and in the Solferino Bridge (Marc Mimram, 1999), over the Seine in Paris. The new movement and dynamics calculations were made using nonlinear analysis as well as wind-tunnel tests for wind load and torsion. The testing determined the need for three tuned mass dampers (TMDs) in the main structure and four viscous dampers at the abutments to respond to dynamic conditions. RFR's work on this bridge furthered the engineering profession's innovations in the area of the dynamic systems for footbridges generally.[6]

These modifications were incorporated in October 2004, and the construction firms of Eiffel, Solétanche Bachy Group, and Joseph Paris began work. The construction of the bridge was an event in itself. The 500-ton steel "lens," or center section of the bridge, fabricated in Lauterbourg on the

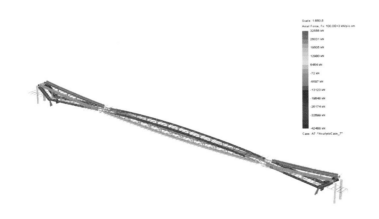

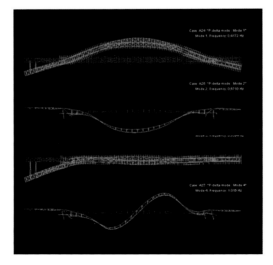

ABOVE: Structural loading diagram.

RIGHT: Dynamic analysis.

OPPOSITE: Models of aeroelastic stability; completed bridge.

ABOVE: "Console" steel framing (left); structural node of arc (right).

BELOW: "Plaza" on the deck of the footbridge.

OPPOSITE: Lens structure transported on the Seine.

Rhine, was shipped up the Seine in the middle of the night and then lifted up on two consoles on each riverbank. The consoles behaved as cantilevers and were removed once the steel structure was erected. Thin bars serve to act as straps and resolve the thrust at the embankment by bending down to anchor the compression bearings at a depth of 18 meters, with 5,500 tons of steel per embankment, and 16 prestressed steel cables anchored, each 30 meters down, holding the tension members in place. The arc is anchored in a block of reinforced concrete. The contractor had to use imported compact material because of a lack of space and the low bearing capacity of the banks. Each arch is a box section made up of steel plates 25 to 80 millimeters thick, and each catenary arc is 1 meter wide and 100 millimeters to 150 millimeters thick. Steel nodes weighing between 5 and 10 tons were cast for placement following a complex geometry and to correct the stress paths at key points of the structure. At the end of the construction process, two small access links as well as the handrails, wood deck, a glass-enclosed elevator, and dynamic devices were installed.

The laminated wood pathways on the bridge give pedestrians the freedom to go up to the street level or down toward the water. At the center of the bridge, where the arc and catenary form a lens shape to stabilize the passage, a unique public place is formed, suspended above the middle of the river and providing a view of Paris. Below, a 65-by-12-meter platform serves as a rest area and a multifunctional urban event space for kiosks and temporary installations such as open-air markets. While in the tradition of many historic Parisian bridges, Bercy-Tolbiac demonstrates Peter Rice's belief that after all of the rules and regulations have been satisfied, projects can, "if intelligently interpreted, have a lot of scope for invention and innovation," through inventive combinations of structural systems and materials.[7]

AVIGNON TRAIN STATION
AVIGNON, FRANCE

The TGV Station in Avignon was part of a comprehensive plan for three new Mediterranean stations in France by the end of the twentieth century. RFR worked with the TGV and SNCF and AREP Architects to design a public space shaped by site and structure.

The site is exposed to harsh southern sun and the mistral, the strong winds of the region, and it is also an earthquake and flood zone. The need for sun shading, ventilation, and wind protection directed the project, along with earthquake- and flood-protection measures; the building is elevated 7 meters on an embankment in an adaptation to the topography. The minimalist design is a new interpretation of nineteenth-century steel-and-glass train sheds. Its narrow, curving form slinks along the tracks and is just wide enough for the passenger platforms.

Two walls, north and south, taper to a center pointed arch, forming the main volume for the 400-meter-long station. The larger, laminated-glass north facade, protected against light and wind by the glass's ceramic fritting, forms the backdrop for departures to Paris, which constitute 80 percent of the traffic. The south side, for arrivals, is a smaller, simpler area shielded from the sun by a concrete wall. The interior's simple horizontal wood-plank panels articulate the space at an elevated platform reached by internal stairs and at a ramp leading to the departures platform. The building's width diminishes as the platform comes to an end, but, rather than terminating with a solid wall, the station opens out to the landscape with laminated-glass sliding doors.

RFR responded to the environmental situation with laminated double-glazed panels that overlap, covering the steel framework that forms the superstructure for both the opaque and the transparent skins of the station. Regularly spaced piles in turn support the triangular frame. The building's shape and the high thermal and spectrophotometric performance of the materials help maintain the climate inside the station. The laminated-glass construction substantially reduces the need for air conditioning, so that the station requires only fresh air, using less energy. Numerous computer models were used to optimize the design and create a form that was both elegant and efficient.

LEFT: Aerial view and south facade.

OPPOSITE: Station concourse (top); north platform (below, right); section (below left).

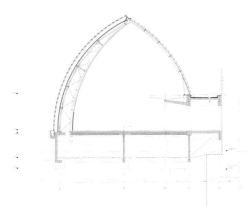

SASAKI AND PARTNERS

Mutsuro Sasaki's prodigious body of work on flux structures, shape design, and the structural potential of fluid space and form have led him down a new path: an investigation of the principles of evolution of living creatures as a way to define and devise structure. This work parallels the scientific foundation of zoologist mathmatician D'Arcy Thompson's project on form and natural structures and the philosophical discourse of philosopher Gilles Deleuze. Sasaki is enhancing computer programs to devise systems for nonlinear structures to self-organize,[1] aspiring to eliminate the hierarchy and distinction between structural and nonstructural elements. In proposing rational solutions based on the mechanical behavior of materials that give physical form to the designs of architects such as Kazuyo Sejima and Ryue Nishizawa of SANAA, Toyo Ito, and Arata Isosaki while fulfilling the safety and stability requirements for withstanding the forces of nature, he has sculpted a new fluid, smooth space.

Sasaki was inspired to study engineering when he was in high school, after seeing the National Indoor Gymnasiums, in Yoyogi in Tokyo (Kenzo Tange, 1964)—a masterpiece in its synthesis of structure, systems, and form. Engineered by Yoshikatsu Tsuboi, the structural elements supporting the organic form are not simple suspension cables but semirigid I-beams that inspired innovation in potential for structural beauty. Sasaki studied both architecture and engineering at Nagoya University, finishing in 1970, after which he worked for Kimura Structural Engineers. He founded his own firm, Sasaki Structural Consultants, in 1980 and received his Ph.D. in 1999, becoming a professor, first at Nagoya and then at Hosei University; he has written numerous essays on structural calculation, structural mechanics, and new computer analysis methods, as well as a book, *Flux Structure*[2] about his structural-design methods and specific projects. Sasaki expanded his firm in 2002 to form a second firm, Sasaki and Partners, with Masahiro Ikeda and Hiroki Kume, but all ventures are managed through the main office.

Sasaki defines the basic design issues before a member of his staff is put in charge of a job. As a participant in the architectural design and form-making process, he expands each project's structural performative potential by exploring specific typologies of structure, pushing for research, and making clear to architects how essential it is that they articulate what they want so that he can find the most appropriate structural solution. This focus on the outcome has led to unexpected possibilities, transforming the architect's design concepts. Sasaki has found that continual feedback between engineer and architect can help maintain the equilibrium between poetic architectural images and rational structural systems. "Concurrence emerges from joint work based on a relationship of trust and mutual understanding while trying to find a solution that simultaneously satisfies the demands of a technician and an artist. This is the moment of greatest tension in the planning stage."[3]

Sasaki begins with a practical, or "structures 101," design process that is extrapolated to a creative, interactive process in which the architect first talks about the project. "It is always like this, whether with a competition when elements are not defined, or with a particular project; when the architect has an idea we have a discussion directly. All the ideas of the projects have to be discussed from the structural standpoint because the idea of structural engineering influences the design itself."[4]

Sasaki begins by developing a hypothetical structural frame that accommodates the building's function. Often involved in the entire building process, he may analyze the stress and deformation generated in the structure by the anticipated mechanical behavior, design cross sections based on the analysis, confirm structural safety, produce construction documents to accurately transmit the results to contractors and third parties, and supervise the site. Sasaki believes the process of developing hypotheses regarding a structure's shape, system, materials, and dimensions is extremely creative: "The structural engineer's role is thereby defined as being more like an architect than an engineer . . . They must undertake their collaborations from an equivalent standing."[5]

For his inspiration, Sasaki looks back to historic buildings in the development of structure with a rational mechanical basis from the domes of Rome and the post-and-lintel frames of Greece, to the concrete shells of Eduardo Torroja, the dynamic hyperbolic parabolic shells of Felix Candela, the inverted hanging shells of Antonio Gaudí, and the thin shell structures of Heinz Isler.

In 1995 Sasaki began working with Sejima and Nishizawa on the Multi-Media Workshop, a curvilinear underground building. He dismissed the architects' original concepts and proposed instead a simple conoid curve similar to the roof of the Sagrada Familia School in Barcelona. The first rendering included steel ribs with steel plates, but the architects wanted holes in it, which would have made the steel frame too light. Also, the water table is high, so Sasaki saw thick concrete construction as being the efficient way to anchor the building, thereby changing the direction of the design.

When analyzing Sejima's structural suggestions, Sasaki asks, "What do you really want to do?" because the end result is about the impression to be made, not about form or shape.[6] For the Ogasawara Museum, or O Museum, in Nagano (1995–99), Sejima envisioned a building that would float above the ground, and Sasaki solved the structural problem by replacing the principal beam direction in a large-span structure with an orthogonal structure hidden on all except the first floor. For an office building in Ushiku, Japan,

RIGHT AND OPPOSITE: O Museum, SANAA, Nagano, Japan, 1995–99.

PAGE 166: Sendai Mediatheque, Toyo Ito, Tokyo, Japan, 2001.

Sasaki placed the cross braces in an "eccentric" position in the core so only the frame was visible. The core balanced the building, and the third floor slab was made thicker so the force could find a path to the brace below. This resulted in the frame being exposed and the glass louvers disappearing at night. In addition, the brackets that support the louvers are cantilevered in an independent structure that can move with the building if necessary, satisfying seismic code requirements. Nishizawa notes, "There is a symbolic message in this solution: although they are often seen as being separate, structure and architecture are becoming integrated."[7] This is a significant aspect of their continuing work with Sasaki, who emphasizes that every situation is atypical and, accordingly, he goes beyond the standard solution.

Sejima presented similar challenges to Sasaki with the design of the Park Café in Koga, where her goal was to achieve a light structure that suggested a trellis of wisteria in the wooded environment. Sasaki's 3-centimeter thick roof and 6-centimeter diameter columns pushed the limits of thinness, immateriality, and transparency of form: "When you reach this point, you create an abstract world where one can no longer distinguish structure from nonstructure—where there's no visible hierarchy anymore."[8]

The Sendai Mediatheque (Toyo Ito, 2000) was a transformative project for Sasaki, inspiring a "more free and relaxed attitude."[9] The six-story media center, with its transparent facade and unique structural composition, was the ultimate structural model for a multilayered architecture, developed in an extrapolation of a Le Corbusier–type Dom-Ino structure and composed of seven steel floor plates supported by thirteen tubular columns. The project began in 1995, with a sketch by Ito, sent to Sasaki by fax from the airport, that showed gridded tubes, swaying like seaweed, with various types of infill and densities, supporting thin horizontal floor plates. Over a period of ten days, the two men completed a series of drawings sent back and forth in a notebook,[10] and from these Sasaki devised a minimalist system of floor plates and tubes that support the plates but that are simultaneously free and united with an overall interdependent structural stability. Using computer simulation and Finite Element analysis, Sasaki could determine structural forms, members, and basic details and establish design criteria for technological objectives that could be mechanically verified. "The end result and the initial shape are close, but to keep their initial shape and image we introduced new ideas such as using steel plates to keep the slab very flat." Sasaki's lightweight floor slab consisted of thin steel sheets with hollow sandwich-plate construction, 40 centimeters thick, and internal ribs spaced 1 meter apart, following the line of stress.

As Sanford Kwinter described, "The Sendai Mediatheque is a supple solid sustained by a saturating system of tensions distributed like pulses and waves across every surface and through its entire volume."[11] Each of the building's three fundamental elements—tube, plate, skin—reflects this common morphogenetic origin. Kwinter notes that the vortexes of the holes within the plates in which the columns support the floor beam maintain a surface tension, which relates to Sasaki's concepts for subsequent work with concrete, free-form shells. The spiral lattice columns are a meshwork of cantilevered steel and function as the service cores efficiently structured in hyperbolic paraboloid curves that resist buckling. The four corner tubes are cantilevered columns and are pin-jointed at their connections above; the top plate rests on a ring beam and connects at each floor, allowing for a loose fit, allowing

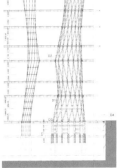

ABOVE: Sendai Mediatheque, Toyo Ito, Sendai, Japan, 2001. Diagram and construction of the columns.

OPPOSITE: Competition scheme for the new Florence Train Station, Arato Isosaki, 2002.

for movement during construction, and avoiding the transfer of bending moment from the floor structure to the steel members. Structural flexibility is essential for movement needed during earthquakes, and the canting of the columns, while seemingly random, improves structural efficiency. Hollow cross sections provide additional structural support at midspan. Sasaki compares the ribs of the steel plates to the more delicate ones in tree leaves and dragonfly wings that correspond to the lines of stress.[12] With Ito, Sasaki explored issues of fluidity, transparency, and nature-based form in architecture.

During his design process, Sasaki employs the computer to analyze a design but not to choose the aesthetic. "Computer programs and design methods for dealing with surface is sensitivity analysis, the one based on evolutionary structure deals with shape. We repeat the calculations to find the solution. It is difficult to predict which solution should be the final result because it is nonlinear. The computer method controls the process that leads to a solution and the result."[13] Or, alternatively, one solution is dictated by the architects' requirement, or by Sasaki, and with the computer programs they control the nonlinear behavior for the structural solution. What the structure will look like is a matter of preference; each solution presented is structurally rational. Sasaki watches the computer monitor during the Finite Element analysis, and he controls the design according to his inspiration based on his own experience and preferences.

Sasaki's design proclivity is toward nonlinear forms, but though the term is common, "nonlinear" does not express a predictable event. "An example is a tensile structure, such as membrane, that can cause an unpredictable deflection or deformation. We used the word 'nonlinear' a long time ago, so it is not a new word, but many people use it now to define what they think is a new geometry."[14]

For the 2002 competition for the new Florence Station, Sasaki, with Arata Isozaki, relied on the three-dimensional Extended Evolutionary Structure Optimization (ESO) method to define forms. The design for the 400-by-40-by-20-meter building, or "flux structure" as described by Isozaki, achieved maximum mechanical behavior with a minimum use of materials while meeting the given design parameters. The structure finds its logic in the evolution of plant shapes toward uniform stress creating a biomorphic treelike branching structure that supports the station's roof.

The ESO method originally was based on eliminating inefficient parts of a structure. In Sasaki's application, he also introduces three-dimensional iso-surfaces (isolines) and bidirectional evolution dependent on the surrounding field of forces. "An efficient and rational structural shape may be sought with a bidirectional evolution that permits restoration and growth . . . By means of the repetitive nonlinear analysis procedure it becomes possible to organically

comprehend the evolution of structural form in the overall structure from the relationships between its shape and mechanical behavior."[15] This shape-design process was then used to create an organically shaped bridge truss-work that supports a steel roof plate that functions similarly to the branching of tree roots. With this system, "it is possible to generate utterly unprecedented structural shapes, heretofore unseen and unimaginable, much like artificial life."[16]

Sasaki sees structural design as moving toward the creation of new three-dimensional architectural structures that possess free, complex, mutable, fluid, and organic characteristics, liberating architecture and opening up the field of engineering. However, in order to implement these in a truly rational way, traditional empirically based structural design methods must be replaced with mathematically based shape-design methods that unify mechanics (rationality) and aesthetics (sensibility). Sasaki has harnessed the Sensitivity Analysis and the Extended ESO methods to generate, via computer, rational structural shapes that apply the principles of evolution and self-organization of living creatures to the making of free-curved surfaces and flux structures that merge structure and aesthetics.

For Sasaki, the key issue about structural feasibility is this: "If there is a project which is possible in terms of engineering but it doesn't have any meaning or interest, it is impossible to make. If there is a project that doesn't have any meaning, then it is impossible for the architecture too." With Sendai, he "found possibility in the architecture" and from that starting point set about finding an engineering solution. Because, although "the first sketch by the architect has nothing to do with mechanical behavioral issues and is not rational," it is where the engineering process begins.

I PROJECT
FUKUOKA, JAPAN

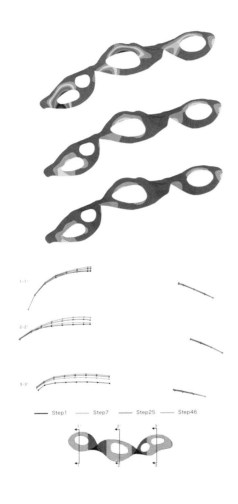

ABOVE AND RIGHT: Structural diagrams showing the stresses of materials and form finding.

OPPOSITE: Concrete shell roof.

The 2005 National Urban Greenery Fair, held on the artificial Island City in Hakata Bay, Fukuoka, includes a 15.3-hectare Central Park and a 1.7-kilometer-long Green Belt, for which Toyo Ito designed a 5,000-square-meter greenhouse pavilion. In a new interpretation of the greenhouse in the tradition of the Mannheim Bundesgartenschau Multihalle (Government Garden Festival; Frei Otto with Ted Happold, 1975) or the Eden Project (Nicholas Grimshaw, 2001), Ito included three 1,000-square-meter greenhouses, each housing a different plant theme. The pavilion's other public spaces—such as reading rooms, workshops, lunch rooms, and cafés—offer visitors educational and gathering places that are linked by fluid spaces. Working with landscape architects Sougo Sekkei Kenkyujo, Ito developed the concept of a circular pattern, like ripples on water, spreading out from the park over the island. These ripples take the form of recessed craters and mounds that mold a new topography for the site, further merging the building and landscape in the form of a planted roof and pathways linking the visitor spaces.

To structurally resolve Ito's undulating, fluid form and still meet the requirements of the structural mechanical behavior and design parameters, Sasaki employed nonlinear Sensitivity Analysis—a method he first used with Arata Isozaki's Kitagata Community Center, in Gifu (2001). Through this method of analysis, Sasaki was able to generate the optimum structural shape while minimizing bending stresses, strain energy, and deformation so that cracks did not result and the rigidity of the shell is maintained. The engineer and the architect, who had worked together previously on the Sendai Mediatheque, applied incremental changes in bending stress and warping, cooperatively obtaining the desired effect. The resulting intensively calculated structure is 190 meters long, has a maximum width of 50 meters, and is 40 centimeters thick. The complex structural shape has a topographic continuity between the inside and the outside, its undulating form reversing its structural flow at two points. The helical, free-form reinforced-concrete shell is similar to the work of Felix Candela for Bacardi Rum (Coyotapec, Mexico, 1960) or that of Eduardo Torroja for the Market Hall (Algeciras, Spain, 1933).

From the initial design, Sasaki calculated the compression and tension forces that have the ability to "transmit loads mainly by axial forces (and very little by bending moments) resulting in the most efficient load transmission, which

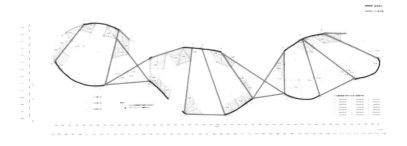

minimizes strain energy in the overall structure."[17] Another "fundamental point is that to analyze structure there should be a form. Since the computer technology is well-developed, you can analyze any kind of complicated shape, but there should be a shape."

Because visitors to the center can walk on the partially planted roof, Sasaki assigned a vertical load of 1.5 tons per square meter as the basis. Some spaces in the roof are open, and the shell twists so the deformation gives a maximum displacement of 2 centimeters, reduced from an initial 14 centimeters. As Sasaki has noted: "In engineering terms, the shell is judged to have converged on a problem-free structure" so that there would be no displacement of the shell. To select the resulting shape, the shape changes at each cross section. He calls the initial shape an "insufficiently rational shape." Sasaki then regularized the shape so that it would be a rigid shell. Computer analysis provides a way for the transformation of each part to be synchronized with the overall structural behavior of the shape and is controlled. It can be compared with a piece of fabric that moves to spread and stretch in tension but then is stiffened for stability and captured, midmotion, at its most efficient form. The building is both sculptural and plastic, pliable but solid, as the essence of the materiality of concrete parallels that of the potential of membrane structures.

As Sasaki was working on this computer model, he posed a question: How would the concrete slab respond if it were thinner, say by 15 centimeters? This, he found, would be "like a piece of grilled bacon on a frying pan, leaping about nine meters at the first step, with a maximum displacement of 70 centimeters. It basically jumped out of the computer screen, in a shape that was infeasible."[18] This reinforced his own thinking, in line with Ito's, that the shapes would be too thin and that the aesthetics and the intuitive structural concepts needed to be developed in concert with each other. The 40-centimeter thickness they eventually arrived at allows a degree of bending stress necessary for

ABOVE: Completed building.

OPPOSITE: Construction with the plywood panels to mold the formwork (left) and concrete combined with the intensive rebar structure (right).

stability. Starting with his own predictions, Sasaki could see the values converge on a specific target and lead to the desired shape and architectural form. Sasaki compares the nonlinear analysis to that of "self-organization, evolution, and artificial life, in that it theoretically generates within a computer structural shapes possessing a sort of biological adaptability."[19] The computer analysis was then output to a laser cutter in Toyo Ito's office to make a physical model. Sasaki's firm was also intimately involved in the intense construction process that the I Project formwork entailed. A complex double formwork of plywood and steel mesh resulted in a smooth reinforced-concrete shell that expressed the plasticity of the form. One idea that Sasaki was not able to put into practice because of high cost was to combine the steel plates as formwork with the concrete in a composite cross section that would allow the formwork to be integrated with the building construction.

ÉCOLE POLYTECHNIQUE FÉDÉRALE DE LAUSANNE, LEARNING CENTER (EPFL)

LAUSANNE, SWITZERLAND

The Lausanne Learning Center designed by architects Kazuyo Sejima + Ryue Nishizawa/SANAA was won in an invited competition in 2004. A unique hybrid structure that combines concrete and steel continues the trajectory of work of the Sasaki office in plate structures. Here each slab is a different system: an undulating concrete floor with a steel roof garden, punctuated with large openings that allow in light and link the levels.

The Learning Center's central entry leads to bubbles containing various programmatic uses, such as a library, language center, multipurpose hall, café, offices, and restaurant. A large undulating single-volume space rather than multiple and separate rooms, the center's levels and ramps are fluid, and its floor and ceiling rise and fall together. Light penetrates the interior through variously sized ovular and round light wells punched in the roof. The 175.5-by-121.5-meter main space is reached from a central entrance hall and around the slab's perimeter. A walkway leads up and down to each of the program spaces, and up to light courts and an inhabitable roof space for a view out.

The roof is supported by the floor structure below, which is a reinforced-concrete free-form shell over the site. The gently curving shape, imagined by SANAA to be somewhat like rolling up the surface of the ground, was given structural rationality through Sasaki's shape-design (Sensitivity Analysis) method. When SANAA first approached him about creating a structure with a span of 200 meters, he determined that this was possible only with a shell.

Sasaki's shell limited load transmission by bending moment; a hollow slab reduced the weight of the shell except at the foot, where it is necessary to resist large axial supporting forces. This main steel column grid of 9 by 9 meters supports the curved main steel beams. Below ground, the flat, reinforced concrete slab was designed with wall columns that resist vertical forces where the shell meets the ground. Columns on a 17-by-10-meter grid support the 60-centimeter-thick slab. The slab also resists the horizontal thrust of the shell as a tie. The German engineers Bollinger + Grohmann are completing the project through the construction in 2008, making precise calculations for the predetermined free-form shell.[20]

Working closely with SANAA, Sasaki used Sensitivity Analysis and evolutionary shape design with Finite Element analysis to create an efficient, mechanically functional structure, one characterized by gradations of light and enclosure and blurring between zones of activities. In the process, they formed a new "shape space" to create a new spatial experience.

RIGHT: Structural analysis models including axial force, bending moment, and vertical deflection as determined in the Finite Element analysis.

OPPOSITE: Plan and model of EPFL.

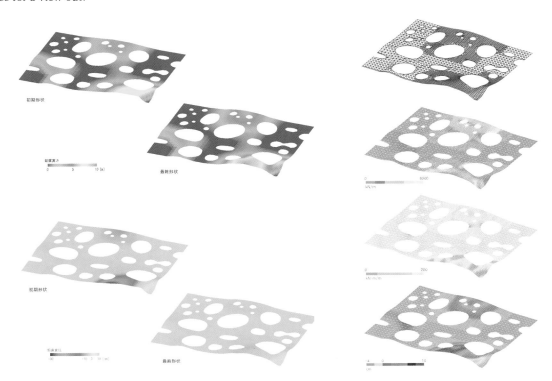

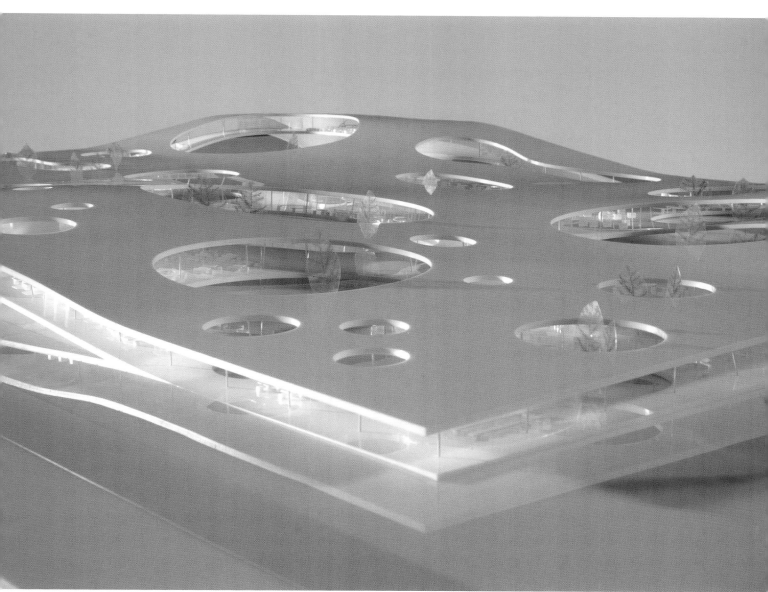

SCHLAICH BERGERMANN UND PARTNER

The engineering partnership of Jörg Schlaich and Rudolf Bergermann, founded in Stuttgart, Germany, in 1980, has grown into a seventy-person office handling projects that range from solar fields and radio towers to bridges and buildings. Some of the firm's work, with partners Andreas Keil, Knut Göppert, Mike Schlaich, and Sven Plieninger heading the practice today, is strictly in the realm of engineering design while other projects have involved close collaboration with architectural firms such as Gehry Partners and von Gerkan, Marg und Partner. The firm's philosophy, which includes social responsibility, has influenced the holistic approach taken by a new generation of engineers.

The use of glass cable-net systems, work in tensile fabric structures, and bridge designs have created an identity for the firm. It could almost be said that the engineers have a stylistic awareness in their structures. But while the firm experiments with a broad range of materials, the conceptual focus is not only on efficiency and economy—an engineer's mantra in the modern era—but also on quality and the poetics of lightness.[1] Experimental work or a new invention deriving from research, technical expertise, and experience in the field might lead to a new formal device that minimizes the structural intensity or doubles the function of a new bridge. But it is the ethical concern—whether about the environmental impact of the materials used, how the public will interact with a structure, understand the structure, or economies of scale—behind the understanding of a structure's impact on society that characterizes the firm's work and clarifies its vision for the structure. Jörg Schlaich refers to the structure that results from this kind of efficiency as a "clean" structure, one that "is respectful of materials, is efficient with minimum effort, and has an understandable structural behavior."

ABOVE: Museum of Hamburg History, von Gerkan Marg und Partner, Hamburg, Germany, 1989. Glass roof and detail of steel connector; DZ Bank, Frank Gehry, Berlin, Germany, 2000. Courtyard roof.

OPPOSITE: Milan Trade Fair, Massimiliano Fuksas, Milan, Italy, 2005. *Vela* glass structure and steel roof (top left); Max-Eyth-See pedestrian bridge, Stuttgart, Germany, 1989 (top right); Hooghly Bridge, Calcutta, India, 1993 (bottom).

PAGE 178: Gahlensche Street Bridge, Bochum, Germany, 2003.

Schlaich and Bergermann both began their careers in the engineering practice of Leonhardt, Andrä and Partners, designing tensile roofs for the 1972 Munich Olympic Park complex (Behnisch + Partner with Frei Otto). Later, Schlaich would further pursue methods for collaborations between engineers and architects. After Otto's appointment as director of the Light Weight Structures Institute at the University of Stuttgart,[2] Schlaich began teaching design to engineers, which he continued through 2000. Otto's influence was also felt in his philosophy of humanism as an engineer. After the Munich Olympics project, Schlaich continued the development of concepts for glass in grid-shell structural systems with the AQUAtoll Swimming Center in Neckarsulm (Kohlmeier and Bechler, 1989) and the Museum of Hamburg History (von Gerkan, Marg und Partner, 1989). Schlaich drew upon similar structures designed by Frei Otto and then expanded the geometric possibilities with new construction and design methods. In designing free-form shell structures, especially those with glass roofs, Schlaich refers to objects and phenomena, such as the mesh of a tennis racket, or that of a kitchen sieve, which when flat can be bent to different shapes, to explain his structures. The mesh has diagonal cables for prestressing through the joints and forms an even length netting.[3] The light, transparent barrel-vaulted dome over the courtyard of the Museum of Hamburg History was created from square glass panes in a steel grid that are then reshaped into rhombi with alternating angles for the transitional free-form dome. The rectangular meshes are diagonally stiffened by prestressed stainless steel end-to-end double cables clamped with joints to transform the grid into a shell. The glazed roof was additionally stabilized with three prestressed and guyed fans similar to those used as spoke wheels to stiffen the Cable Net Cooling Tower in Schmehausen (1974).

Schlaich and Hans Schober, who heads the firm's New York office, have also designed a free-form gridded glass roof for Frank Gehry, who was interested in achieving irregular geometric forms for the DZ Bank in Berlin (2000), with triangular glass panels and stainless-steel profiles in a double-curved surface to cover the 61-by-20-meter atrium. For the glass roof of Gehry's Museum of Tolerance in Jerusalem (2006), Schlaich and Schober developed a method in which the free-form glass roof follows a logical geometrical approach. They used a quadrangular mesh stiffened by diagonal cables, allowing the four corners of the mesh to be in one plane. The lengths of the individual elements vary harmonically, as do the angles of the mesh. In both projects, the two engineers grasped Gehry's visual and formal intentions and translated them into structural logic.

The most expressive free-form glass roof that Schlaich Bergermann und Partner has undertaken was for Massimiliano Fuksas's Trade Fair complex in Milan (2005). However, for the engineers, the project goes beyond the line of efficiency and economy, formulating the next paradigm in design engineering, but also begs the question whether the expressiveness is necessary.[4] The complex glass and steel roof covers the 1,300-meter-long and 30-meter-wide passageway between the exhibition pavilions.

Fuksas calls the undulating forms the *vela* (sail) and compares it to the Alps when viewed from a distance. The roof height varies from 16 meters to more than 37 meters at the cone-shaped protrusion over the entrance called the *logo*, as it flows below and above its own primary surface: downwards like a black hole in space and upwards like a frozen volcanic eruption as a liquid solid, suspended in time and space. The mesh structure of the logo-roof is a combination of two helixical nets and a net with four specific points to carry the transitional stresses between the curve types. The roof is veil-like in its

translucency, 50 meters in diameter, and is supported by 183 steel columns, which then split to smaller steel supporting treelike arms, above the 12-meter height. The undulating surfaces, captured in a glass mesh megastructure, create a new type of light structure striving towards a fluid shape space.[5]

Schlaich and Bergermann have also exploited the form-making potential of tensile fabric and air-inflated tensile roofs, especially for large-span sports complexes and temporary structures. Expanding on concepts developed by Otto and those of Horst Berger, they have used both physical models and computer programs for their exploration of the form-making possibilities of flexible kinetic structures prestressed by means of cables at key locations. The roof for the Roman amphitheater in Nîmes, France (Labfac: Finn Geiple, Nicolas Michelin, Architects, 1988), was one of his first applications of air-inflated fabric, which allowed use of the building in the winter months. With a 1-millimeter-thick PVC-coated polyester tensile fabric cushioned in a 60-by-90-meter prestressed membrane, they inflated the membrane to form a dome shape. It was stabilized with a polygonal compression ring, recessed into the elliptical arena, and restrained by a cable net. A similar system was devised for the Bullring in Madrid (2000), where Mike Schlaich opted for an inflated ETFE foil cushion membrane, achieving a more transparent roof.

Of bridges being built today, Schlaich Bergermann's are among the most artful in their structure and form. The firm has designed highway and pedestrian bridges around the world so that locally available technologies, materials, and craftspeople are part of the construction process. Schlaich emphasizes the joy of engineering generally, but feels a particular connection to the design of footbridges. These are bridges that "we touch, and therefore it is very important that they have human scale . . . A footbridge has a wide field of possibilities, and it is a lot of fun insofar as the quest of a footbridge does not need as much structural force as for a highway or railway bridge, and therefore can accept some additional design concepts." They also exemplify public-works construction and urbanism and, for Schlaich, are an expression of concern for civic life: "The public and clients have to realize that a bridge has just as much impact on our surroundings as a museum or other large building, and thus needs to be as carefully considered."

Two of the firm's curvilinear suspension footbridges—the Gahlensche Street Bridge (Bochum, 2003) and the Max-Eyth-See pedestrian bridge (Stuttgart, 1989)—have specific design engineering characteristics. In the Bochum bridge, the S-curve of the deck is supported by two inclined masts inside the curves maintaining the floating qualities of the elevated snakelike form. The masts are anchored by the cables that they bear, rather than with guy wires in what is called a "cable trick." The curving of the deck acts to stiffen it against the pull on the cables helping to stabilize the mast. The trick is the result of cables stabilizing the mast from the deck rather than going to the ground.[6] The two curves also work to brace each other all in a complex interaction of forces. The firm explored the load-bearing properties of the circular girder with their Rhine Main Danube Canal Bridge (Ackermann + Partner, 1987), and this led to the idea of using the cables in tension and struts in compression, one above and the other below the walkway slab.

Schlaich Bergermann und Partner has developed complex folding footbridges, such as the bridge at Kiel (1997), which relies on an integral movement mechanism to pull the platform into a fold and swivel out of the path of boats entering or leaving the harbor. The challenge for the Humpback Footbridge across the Inner Harbor at Duisburg (1999) was to open occasionally for large boats. This need was met by hinging the decking elements. Extra planks in the abutments meet the arched bridge, allowing the masts to tilt outward, at which point the back stays are shortened, and the main cable flattens out to lift the deck in an arched, or humpback, shape in three preset positions up to 9.2 meters in height. Surprisingly, even when the bridge is arched in the middle position, it can still be walked on.

In collaboration with Feichtinger Architectes, Andreas Keil designed a pedestrian bridge opening in 2008 that will carry three million tourists a year (and the occasional car) from the mainland of France to the historic island of Mont-Saint-Michel. This minimalist, low-lying structure, similar to the Rapperswil Pedestrian Bridge over Lake Zurich (Walter Beiler, 2003), is constructed of prestressed concrete on steel supports and finished with wooden decking. The minimal profile will only lightly impact the landscape and the historic site.

The firm is also recognized for designs of communications technology infrastructures such as telecommunications masts and lookout towers. Of note is the Killesberg Tower in Stuttgart (2001), in which a double winding staircase leads to a series of viewing platforms at different levels, suspended in a net of forty-eight cables mounted between compression rings, resulting in a large-scale mesh network and held together on a central, compressed mast. Each level is also stiffened horizontally through the four platforms. The design is similar to the firm's earliest cooling towers casings, such as the one at Schmehausen (1974), a series of hyperbolic curves in a filigree cable network.

Schlaich Bergermann's humanism is expressed by its passion for energy conservation solutions, solar energy production, and projects in developing nations. A special division in its office, managed by Wolfgang Schiel, works on the design of vast fields of solar collectors, as well as updraft towers or solar chimneys, with a solar field platform completed in Almeria, Spain (1990 and on-going), a tower in Manzanares, Spain (1982), and plans for an updraft tower in NSW Mildura Australia (2008). Their work as specialists in this field combines expertise in engineering and energy production.

The firm continues to experiment with structure and to put forward ideas about the beauty of functionalism in both architecture and infrastructure, through projects such as the Novartis building in Basel, Switzerland (Frank Gehry, 2007), the Nam June Paik Museum in Gyonggei, Korea (Schemel Stankovic Architects, 2007), and through investigations into new lightweight materials.

The question of functionalism as beauty fascinates Schlaich: "A flower is not a structure just for the fun of it: it has a shape, color, and behavior that is purely functional—and therefore it is beautiful. But I cannot turn it around. It is not automatically true that if the structure works, then the building is beautiful—then I could ask a computer to do it. There are exceptions to the rule, such as the Sydney Opera House, which is a poor structure but a great building. But it is a very rare case. Usually this will not happen, and the building will quickly lose interest and attraction. The difference between an architectural and an engineering task is its complexity. A bridge has a simple function: it connects two points, but there are an infinite number of ways to do it. Finding the solutions, not just the mechanical processes, is what makes engineering an art and in turn makes our job creative."

BELOW: Prototype Solar Chimney, Manzanares, Spain, 1981.

OPPOSITE: Rendering of bridge to Mont-Saint-Michel, Feichtinger Architectes, France, 2008 (top); Killesburg Tower, double spiral lookout tower stair, Stuttgart, Germany, 2001 (center); Humpback Bridge across the Inner Harbor, Duisberg, Germany, 1999 (below).

HAUPTBAHNHOF (BERLIN CENTRAL RAIL STATION)
BERLIN

The iconic nineteenth-century European train station with its latticework steel trusses, grandiose height, and parabolic roofs, ideal for train movement and the dispersion of steam exhaust, has been revived but adapted for today's needs in the new Hauptbahnhof (Berlin Central Rail Station), formerly the Lehrter Bahnhof, completed in 2006. The new urban complex designed by von Gerkan, Marg und Partner and Schlaich Bergermann und Partner is exceptional both in its technology and form and in that it symbolically and physically links the disparate parts of post-unification Berlin—north/south and east/west—via an infrastructure of mobility.

This 160-meter-long and 40-meter-wide, multilevel, mixed-use urban hub is another of Schlaich Bergermann's deft translations of form to structure including girder and cable supports, the four viaducts, the two-way curve of the glass panes, and the cable facade system. Working from the beginning of the project to resolve the complex program with the architects, partner Hans Schober designed a 450-meter-long barrel-vaulted glass roof (completed only partially to 320 meters) as a three-dimensionally curved shell, which ranges in height from 14.5 to 16.5 meters over the east–west platform. A main arched vault of 66 meters forms the largest span and slopes steeply down to the train platforms. The two-way curvature of the cylindrical vault is formed by the shift of dimensions in the rectangular glass panes, and variations in the steel framework form the mesh roof. The arch exploited the shell curvature to make a light and efficient structure braced with welded T-sections to diagonal cables. The arched basket-handle-shaped frames, spaced 13 meters on center, stiffen the structure in conjunction with the cable support below and above the frame, following the bending-moment diagram to minimize the tension in the steel members of the arch so that it remains in compression.[7]

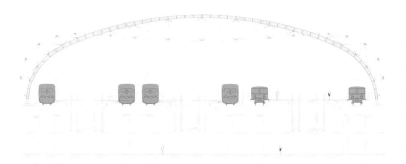

Trains on the east–west lines arrive in the station via the Humboldt Harbor Bridge, completed in 1999, elevated 10 meters above the train station. The busy hub below is visible through the deck of the train bridge, which is supported on slender, treelike

ABOVE: Station platform and the curved shell arch.

OPPOSITE: Section of train shed (above); station interior (below).

ABOVE: View from northwest (left); glass facade (right).

OPPOSITE: Multi-level space above the tracks.

branching steel columns. The main viaduct, an elegant 240-meter-long polygonal arch bridge, spans Humboldt Harbor with a 60-meter-long arched clearance for ships. The prestressed-concrete track deck is supported by 60-centimeter diameter cast-steel tubular columns and 10-centimeter-thick walls help to maintain the narrow profile. This is the first use of cast steel in a railway bridge. Over the train hall's main point of entrance for the east–west trains, 23-meter-high steel columns support the bridge deck in a grouping of four 50.8-centimeter diameter steel tubes, resulting in the branchlike configuration. Cast-steel nodes connect the columns to the underside of the deck. Two glass rectilinear office buildings, clad in a similar way as the train shed, rise over the platforms. The street entrance that leads to the station's retail spaces and to the passenger platforms is covered with a 213-meter-long glass barrel-vaulted roof.

It is interrupted by a 41-meter free-span opening between the office structures but intersects with the roof of the east–west platform roof in a classic groin-vault system, supported on its corners by suspension rods anchored to the office buildings. At each facade, the engineers employed a system of pretensioned cables that run behind vertical glass joints and crisscross at each glass pane. The cables are connected with steel pins to pairs of three-ply laminated glass fins that can support loads of up to 10 metric tons to vertical cables.

Train activity made construction complex so trusses were assembled, rotated, and lowered into place, and sections of the new station were built on the adjacent site and slid into place on rails. The structural proficiency enabled the creation of a new urban three-dimensional space, expressed in layers of materials, form, and function.

HESSENRING FOOTBRIDGE
BAD HOMBURG, GERMANY

Schlaich Bergermann und Partner's design for a cable-stayed footbridge in Bad Homburg—a commission they won in a competition—achieves material strength without sacrificing visual lightness, making the bridge design unique as well as structurally avant-garde. Conscious of creating an urban event along with a new entry to the city, the engineers wove the bridge into the fabric of the site, over a divided highway, by tucking one end of the structure into an existing underground garage and by placing stairs and elevators on supports on the opposite side to reach the city center—encouraging a continuous flow of pedestrian movement. Glass bricks close off the bridge abutments (monolithic on the garage side), and at night the lights in the handrails and on the mast make the form glow.

Knut Göppert designed and organized the construction of the 75-meter-long bridge around an unusual central focal point: a 15-meter-high mast with four treelike branches made of dense stone—homogenous Zimbabwean Gabbro, Nero Assoluto—held in compression with a centric tie and supported at its base on a steel hinge bearing. The mast supports sixteen steel cables that suspend the 46-meter-long decking along a 6.9-meter-wide superstructure and 30-centimeter-thick concrete slab walkway equally distributing the loads.

In a rare heavy, blocky material choice for Göppert, stone was selected for its ability to make double-curved exterior surfaces. The mast is formed by nineteen individual stone blocks, stacked like vertebrae, drilled through to a centric core, and sealed with high-strength mortar. The tension bars are threaded through the stones and prestressed to increase the compression until no tension occurs under any working load. In order to test the manufacturing sequencing, measurements, and final shape, the stonemasons created a full-scale sandstone model.

The Hessenring Footbridge is a sculptural icon using traditional materials, and while physically heavy, its delicate aesthetic and structural poetics speak of ingenuity and civic pride.

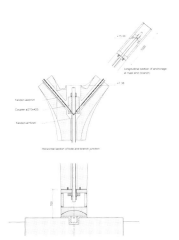  

ABOVE: Diagram of stone pier fabrication (left); pedestrian platform (center); evening view of the bridge linking two parts of the city (right).

OPPOSITE: Central mast support, executed in Zimbabwean stone.

COMMERZBANK ARENA
FRANKFURT, GERMANY

Schlaich Bergermann und Partner has been developing variations on operable tensile roof structures in stadiums since their work on the design for the kinetic roof of the arena in Nîmes, France, in 1988. There the roof is dismounted every six months, with the good weather, and remounted again in a low-lying recession in the arena. Its self-supported air-cushion is held in place by thirty pendulum stanchions encircling the stadium. While that system functions well, for the reconstruction of the 1925 Waldstadion (Forest Stadium) in Frankfurt, Germany, the need was for a more flexible roof that could be opened and closed at any time, adjusting to the weather at a moment's notice.

The new stadium designed by von Gerkan, Marg und Partner was commissioned as the Commerzbank Arena for the 2006 FIFA World Cup Championship, and a capacity for 45,000 spectators was planned. In addition to the challenge of the roof structure, the engineers had to rebuild the new stadium where the pitch was the only relict from the former Waldstadion. The stadium design also featured new media capabilities, space for lounges and private boxes, a rainwater collection system and drainage facility, and an 1,800-car garage and catering spaces situated below the stadium. In addition, the construction sequence had to be coordinated so that local teams could continue to play. Construction began in 2002, and the east and west ends were rebuilt; the main stands were then replaced. The stands were prefabricated in segments that were bolted together on-site, and the seats installed in a flexible system that makes possible their removal as the venue's different events require.

Knut Göppert, the partner on the Schlaich Bergermann team, designed the kinetic membrane

BELOW: Elevations of the stadium and diagrams of the roof in open position.

OPPOSITE: Central fasteners for fabric.

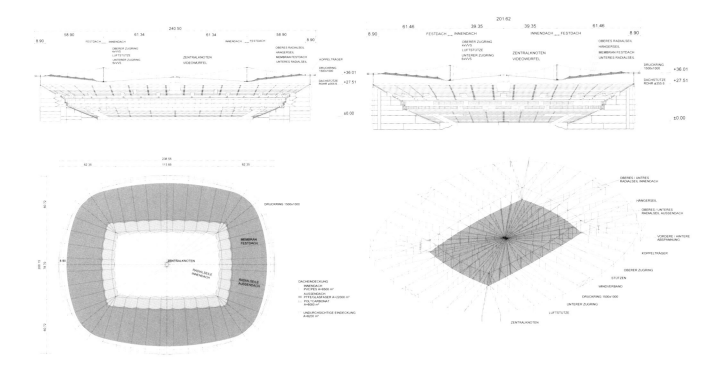

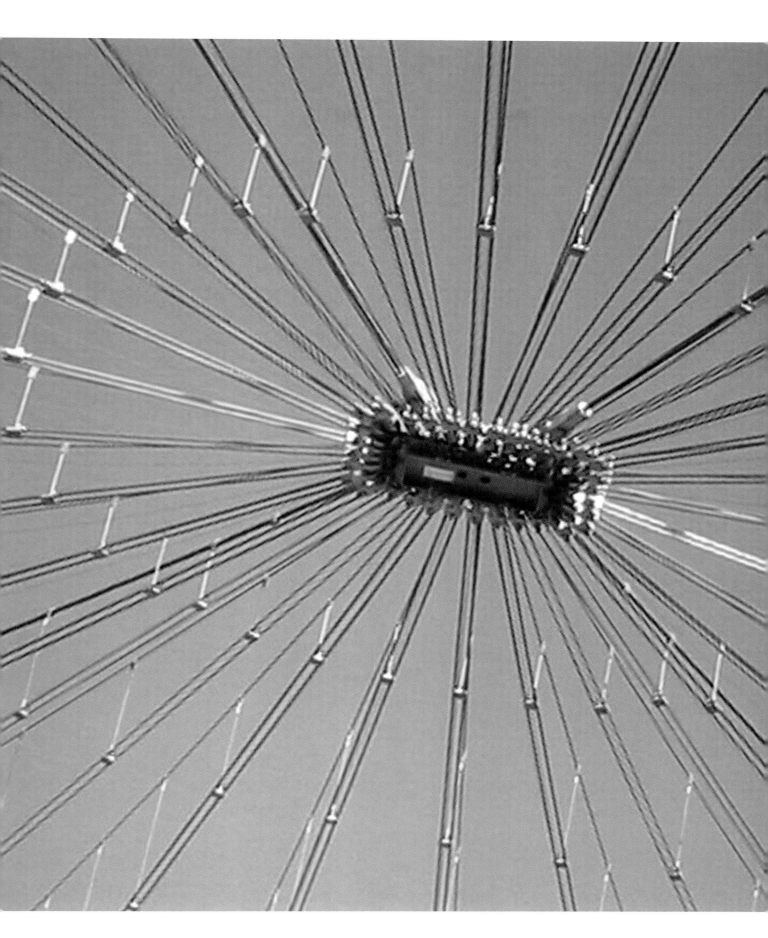

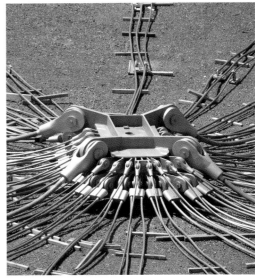

CLOCKWISE FROM LEFT: Roof; cables and fasteners; media cube in position; roof being raised into position.

OPPOSITE: Membrane fabric ready to be raised.

retractable roof system supported by an oval ring with an outer compression and two inner tension rings: 9,000 square meters fold into the center of the tensile cable structure, which gathers itself up via mechanical controls for storage in a 30-ton video-cube. The cube, suspended in the center of the stadium, has screens on each surface and can be moved vertically to offer clear sightlines to the field. The suspension system's steel cable-net structure consists of an upper and lower steel rope net connected at ground level. The engineers orchestrated the tensioning so that it would occur simultaneously on all forty-four axes, and the membrane structure was then installed on the outside of the tensioned ropes. The roof was the largest single membrane panel built at the time.

WERNER SOBEK

Engineer Werner Sobek and architect Helmut Jahn coined the term "Archi-Neering"[1] to express the idea of the collaboration between architects and engineers and the combining of the technical and aesthetic interests of their particular disciplines. This cross-fertilization evolved from Sobek's combined studies in architecture and engineering at the University of Stuttgart, where he received his Ph.D. in 1987, under Jörg Schlaich. In developing his own curriculum, he merged disciplines, studying with aircraft, automotive, and textile designers, learning the basics of "how you weave a pair of jeans, or an astronaut's suit; how fashion relates to fabric as structure; how you forge and caste in the automotive industry; and why they freeze the rivets in the aircraft industry." This preoccupation with how materials and structures are integrated led him to engineering design.

Sobek has developed a comprehensive approach to the resolution of a problem, and the ability to imagine a project from the viewpoint of the architect, engineer, client, and even critic. He attributes his open perspective and search for an honesty of expression at many scales to his roots in the industrial and precision-machine culture of the German Schwaben region, where he was raised and which influenced his desire to design with efficiency, elegance, and concern for sustainability combined with an appreciation for the aesthetics of technology. After working for Schlaich Bergermann und Partner on projects such as the Roman arena in Nîmes, France (Finn Geipel and Nicolas Michelin, 1991), which introduced a temporary pneumatic roof structure over the historic arena, Sobek became the director of the Institute for Structural Design and Building Methods, located in Hannover, Germany (1991–94). He founded his firm in Stuttgart in 1992, with the first project the Rothenbaum Stadium (Schweger & Partner, 1993), which synthesized his interests in fabric and lightweight kinetic structures that incorporate machine components. In 1994 he succeeded Frei Otto as professor at the Institute for Lightweight Structures at the

ABOVE: Installation of the membrane fabric for the Rothenbaum Stadium, Schweger & Partner, Germany, 1993.

OPPOSITE: View from the atrium of Deustch Post Tower, Murphy/Jahn, Bonn, Germany, 1997.

PAGE 194: Deustch Post Tower, Murphy/Jahn, Bonn, Germany, 1997.

University of Stuttgart. In 2000, Werner Sobek merged this institute with the Institute for Construction and Design into the new Institute for Lightweight Structures and Conceptual Design.

As a professor, Sobek has conducted complex research on materials and methods at the Institute for Lightweight Structures, where a glass dome was installed for the purpose of experimenting with using an epoxy resin to glue curved 10-millimeter-thick glass panes to trace changes over time in various climatic conditions. This research applies Frei Otto's studies of soap bubbles and foam to tensile structures. The soap bubble in its structure and form maximizes the coverage of an area with the minimal surface. As Sobek expressed it, "We are interested in how you can influence the size of the bubble and the distribution of those bubbles within a bubble foam substance, where we predict the distribution of the size of the bubbles, clear span areas, or less dense, or light areas. If you could do so, then you could arrange the density in a way that there is optimum light transition and flow." He transfers his research into a real project when "his brain gives the okay." But he tells his students that knowledge "slips slowly to the stomach, and as it gets down to the non-thought level of the brain, and you get the feeling in your fingertips that all the proportions would be okay; only then do I do the building. Intuition as it absorbs scientific knowledge takes time to transfer into the built piece of a building. Goethe said that you can step up over the border and be avant-garde only if you know where the border is."

In Sobek's experiments with tensile materials and kinetic structures the taut fabric skin is no longer his goal; instead, he reveres the way fashion designers drape fabrics. A wrinkle-free tent structure, to him, has no sense of scale or massing, and the shape is fixed. With brick, or masonry, the rhythm of the material creates that sense of scale, and this inspired Sobek's investigation of how to incorporate wrinkles and folds into fabric structures in order to convey a sense of scale. He also sees some ecological potential—in the form of heat storage—in transferring the technology being used to engineer fibers for athletic clothes to building-skin technologies.

Sobek's investigations have often resulted in incremental changes in building details that transform complex structural systems. Referencing the model of the 1972 Munich Olympic Stadium, Sobek's firm applied the configuration of the cable-net system, with its heavy black joints between panes, to the engineering design for the Rhoen Clinic in Bad Neustadt (Lamm, Weber + Donath, 1997). In that use, unusual lightness resulted from the structural configuration, which began with glass panes being clipped to a metal-wire framework similar to the knots in a fisherman's net, then layered like fish scales to form a fluid roof over a winter garden. Because the structure did not have to be completely sealed, the shingles were able to flow with the breeze. This shingle detail clip system enabled an atmospheric change and was lighter than the earlier Munich Stadium.

Material transformations fascinate Sobek, as in the subtle dematerialization of glass brought about with various lighting conditions or actual applied coatings from opaqueness to transparency, or the ephemeral qualities of materials that react and change in different climates. Of facade-engineering projects such as the Kaufhof Chemnitz (Murphy/Jahn, 2001) or the student building at Bremen University (Alsop-Stormer, 2000), he says, "I often seek total transparency, not for its own sake but as an architectural element when combined in opposition with varying degrees of opacity." In Bremen's glass facade, Sobek devised a cable system stretching taut vertical cables 50 feet from the roof to the foundation, eliminating horizontal cables to increase transparency. The

three-quarter-inch prestressed vertical cables had to be anchored with huge stay-spring tension bearings, creating the architectural and environmental effect of a free-floating and frameless facade.

Sobek's interest in glass and his collaboration with architects correspond to his repertoire of sustainable, "high-tech" structures, as exemplified by the Deutsch Post office building, in Bonn (Murphy/Jahn, 2003). This structure is composed of two semielliptical forms forty-one stories high that are linked at four levels by glass wintergarden bridges that span the full-height atrium. Working with Matthias Schuler of Transsolar ClimateEngineering, Sobek and Jahn conceived the steel-grid double-glass facade as a breathable filter, its operable windows allowing air into the internal space, where it cools down and then is distributed to the office spaces. This cooling process is also achieved with water flowing through the coffered reinforced-concrete ceilings—another component of the environmental-control system. As Sobek said: "We can control the building shells. The shells will react in a perfectly natural way to internal and external changes. But this means that they have to be built with a certain degree of technical sophistication that demands considerable engineering input and also has a significant impact upon architectural design."[2] Glass forms a protective boundary without breaking the visual connection between inside and outside, and it is this dichotomy that has led Sobek to a kind of industrial-design purism: as the more technical aspects of buildings become increasingly efficient, every aspect of the design is integrated.

Sustainability is understood as a parameter in the firm's work: "Like beauty, it has to be there, but is not a headline." Thus, climate and systems engineering is now an integral part of the design process, and buildings constructed with recyclable and recycled materials and operated with

renewable energy and no emissions are the new standards.[3] Concerns about the environmental degradation caused by construction waste drive the work toward the ideal of full recycling of an entire building—the steel, plumbing, electrical wiring, copper, PVC pipe, polyfibers, and even paint. The goal is that "the building disappears at the end of its life with honesty without leaving tons of waste."

The new Saint Etienne Cité du Design, France (LIN: Finn Geipel, Giulia Andi, 2007)—a mixed-use facility for research, education, and communication in the field of design—exemplifies the firm's innovations in sustainable design. A former weapons factory has been renovated as workshops and classroom space, and a new building was built within the existing complex to serve as a free-floating body, or "platine" (long platinum circuit board). Sobek completed the structural engineering and used solar panels as a thin cladding that can adapt to the changing programmatic needs in the interior, adjusting amounts of light or air accordingly. The solar panels act as a solar-power plant to generate some of the energy needed for operating and give the building its identity.

Sobek believes that the "qualities" and identities of his buildings "must be readable from the basement to the top floor. That means that if a column is welded to the girder at the ground floor, it must also be at the top floor; the tube has to be the same throughout the building, and all the details are repeated with an inner logic. That is perhaps where style is found, because it must hold true for the entity." This is visible in his interest in designed objects, in his attention to the play of light and shadow, and in his concern for creating beauty. Like a car designer, Sobek uses shadow line to emphasize volume. He considers the impact that structural elements can have on form and how shadow and space-making potential direct a project.

In a provocative urban-design study, Sobek prepared a scheme for a new mixed-use tunnel along the B14 highway to unite the two sides of central Stuttgart. The concept was for a 1.25-mile-long folded lattice structure of annular elements, clad in glass, titanium, and fabric panels whose thermal conductivity and light transmission would respond to environmental conditions. The tunnel's commercial, community, and cultural space would have a new central core and connect with a sunken pedestrian area in an artificial landscape.[4]

As Sobek looks to the future, he envisions manipulating the physical properties of structural members in ways that allow them to adapt to the flow of forces. In the case of a railroad bridge—normally designed for that brief period when its load or the weather conditions it must endure are at their most extreme but which for all other times is overengineered and underloaded—Sobek asks how to artificially soften the structure making a homogenized load level. "Imagine that you make the lightest bridge, introduce activity of a few structural elements in artificial manipulation and by doing so lower the dead weight to 50 to 60 percent and use half as much material. The traditional engineers will say, 'Do you want to do that and put people in danger?' Of course not, I am thinking about how to do it in the future."

BELOW: Rendering of urban design scheme B14 for Stuttgart, Germany.

OPPOSITE: Saint Etienne, Finn Geipel, Giulia Andi, Cité du Design, France, 2007. Interior rendering and model.

AUDI AG EXHIBITION STANDS
FRANKFURT, GERMANY

In a small but complex project, Werner Sobek designed, with Ingenhoven Overdiek und Partner, the structure for a series of flexible exhibition stands for Audi AG, first displayed at the Frankfurt Motor Show in 1999 and then installed at other automotive trade shows internationally. Ingenhoven and Overdiek had imagined a banded, curved glazed structure that, at first, was impossible to realize because of the need to bend the form with its components in three dimensions. But, through the use of a cable-net structural system, Sobek developed a system of interlocking connections and grids that would curve in response to the form.

The Loop, as it was called, was 6 meters high and could vary in length from 100 to 300 meters, depending on the show's requirements. Frosted-glass panes of standard safety glass were secured in a narrow-mesh stainless-steel cable net and, in turn, mounted in tension in a curved, tubular, stainless-steel web structure. The 4-millimeter-thick glass was keyed into the web through the system of knots, or steel beads—5,840 in total—Sobek devised. Although the first calculations specified an overly complex system containing 13,000 different glass panes, such specificity was eliminated in favor of increased mass production, again in a nod to industrial design. Thus, computer numerical control (CNC) milling was used to cut the 300 different types of glass panes into triangles and to shape them in conformance with the stand's overall undulation.

The interior of the stand had oak floors that curved up slightly to meet the curve of the walls. Audi displayed their latest vehicles in five bays as visitors passed in and out of the stands through curvilinear openings. Another stand contained an elliptical staircase that ascended to meeting rooms, a cafeteria, and visitor lounge areas with viewing balconies. Heating, ventilation, and mechanical systems were integrated within the stands, as were the video and lighting systems that relayed the car manufacturer's message. Through standardized assembly of components, similar to that of automobile assembly, the portable display structure was designed to be erected sequentially and quickly in each location. On the curved interior surface, projected light and images created an overall effect of a luminous skin within a delicate silver necklace.

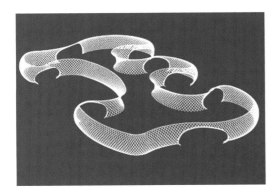

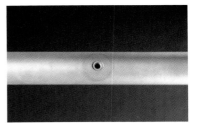
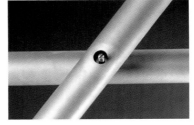

ABOVE: Computer rending of undulating facade structure (left); stainless-steel elements of the meshwork that secures the glass panels, and details of the steel nodes (right).

OPPOSITE: Exhibition stand installed and illuminated.

MERCEDES-BENZ MUSEUM
STUTTGART, GERMANY

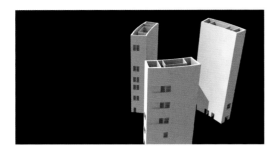

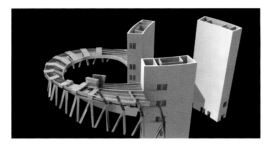

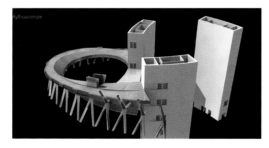

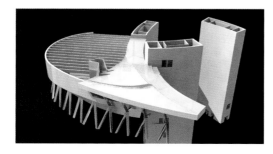

Sited adjacent to an existing facility that had outgrown its space, DaimlerChrysler's new Mercedes-Benz Museum, in Stuttgart, was designed by architects UNStudio (Ben van Berkel and Caroline Bos), of Amsterdam, with Werner Sobek Ingenieure and Transsolar, in a 2001 competition. It opened in 2006.

UNStudio based their design on the trefoil recalling the Mercedes logo. From this point of departure the firm created a complex, reinforced concrete, 47-meter-high double helical building clad in aluminum and glass that is double the size of the Guggenheim Museum (Frank Lloyd Wright, 1959). The fluidity achieved recalls highway cloverleafs with their dynamic interchange of surface, volume, and function. Here they are expressed holistically and enclosed in the 75,000-square-meter space.[5]

The visitor enters through an expansive first floor entry lobby to take a capsulelike elevator up through the open interior atrium to the exhibition areas beginning on the top and ninth floor. From there, one of the two intertwining thematic spirals—the "Legend" or history of the automobile, its associations, and the people connected with it and the other theme, the "Collection," displaying the chronological histories of the vehicles—can be followed down nine levels in two pathways, with crossover spaces on each floor. The trefoil plan is thus combined with a diagram of the double helix loop through circular overlapping of interior spaces formed by Sobek's unique structural system.

Spanning 30 meters of uninterrupted space, each exhibition floor is a leaf of the trefoil, offset from the adjacent floor by a 120 degree rotation and shifting in separate 1-meter-high ramps at each level as they rotate around the central triangular void forming six horizontal plateaus up the building. Beams span from one of the three central cores to a box girder at the perimeter that varies in section from 4.5 meters wide by 7.1 meters high, where it originates at the core, to 35 meters wide by 1.4 meters high, forming an intriguing structure in itself called the "Twist." This box girder's hollow section allows for the installation of all of the mechanical services. The Twist transfers its own load and the load of the floor slabs to the central core and the inclined columns at the facades. "A series of vertical partitions in the hollow part of the girders ensures continuous load transfer between the ramps and the columns." The top section of the Twist forms the ramps and stairs that lead from one level to the next and are situated along the exterior wall, forming oblique pathways, and creating shortcuts between the floors that also allow for long vistas through the building, contributing to a voluminous atmosphere. The slopes in the floors also necessitated a shift in the angles of the exterior glazing as it wraps the building. Sobek's office designed the bearing elements from the three massive concrete cores to the floor slabs, the ramps of the Legend areas, slanted facade columns, and the Twist. As Ben van Berkel said of how the structure enabled the desired space: "I like it when you have a curve that becomes horizontal but that is vertical, and you can look up and down the void space. What is so beautiful about a curved space is that you can see where you come from, since it is a double helix not a spiral. It is not dogmatic;

ABOVE: Atrium.

OPPOSITE: Computer renderings of the contstruction sequence of the three cores and the Twist showing the concrete floor slabs that create the double curves.

CLOCKWISE FROM LEFT: View of the site from construction crane; installation of angled concrete column structure; completed building; concrete formwork.

OPPOSITE: Completed building at night.

you can select to go through one exhibition space or the other from the "Legend" to the "Collection." The most important thing is the freedom to lose your way but always find the entrance. It is like walking in a garden, rambling, meandering around, feeling lost, but always returning to your starting point."[6]

The tetrapodal (four-legged) columns support "bending edges" of the glass curtain wall, and vary in angles and lengths. The facade, in its complex curvature, combines aluminium and glass designed and calculated by Sobek's office with three-dimensional computer modelling, resulting in 2,700 square meters of cladding elements that sit tightly forming a unified skin, like a car body. "During the design phase for the window structure, two options were considered: one with glass braces acting as uprights, and one with thin welded sectional steel. Due to budget considerations, the latter was chosen. Because of the building's notable geometric complexity, about 1,700 different slab formats were required."

A plastic film on the windows with a silk-screen pattern in trapezoidal-shaped panes provides various levels of transparency into the exhibition areas, which furthers the spatial effects and shields direct sunlight. The aluminium sheets are 1.7 by 5 meters and are made of 4-millimeter-thick sheeting, part of which have a spherical curvature, with rounded edges and a layer of Duraflon.

The exposed zigzag of the concrete columns in the interior, combined with the cladding system, speak to an integration of structural design forming a spherical unfolding of the interior space. The difference between open and enclosed spaces is reemphasized by the variations in the facade and focuses attention on the sculptural qualities of the building.

R128 HOUSE
STUTTGART, GERMANY

R128 House, in Stuttgart, exemplifies Werner Sobek's approach to engineering and environmental design, following a set of principles that lead to design solutions that are far-reaching in their research implications. The house, which he both designed and engineered, is a personal manifestation and consolidation of ideas about sustainability and construction. It is a twenty-first-century interpretation of high-tech construction, but it shows that his sensibilities lean more in the direction of ecological functionalism and the hum of buildings as living mechanical organisms.

The four-story glass cube is made of prefabricated, recyclable parts and was assembled on-site in four days. Similar to the new direction in Germany in automotive and electronics recycling, the house is based on an idea of ephemerality. Sobek chose to expose the guts of the building on the interior such that electric cables and plumbing pipes are routed in ducts that can be accessed easily for repairs and upgrades rather than being encased in a solid wall. Each component is marked so it is identifiable throughout the entire building, and components that are removed can be reused, disassembled, or recycled.

The 11.2-meter-high steel frame is bolted at each of the four floors. Twelve columns on a 3.85-by-2.90-meter grid are connected with beams, with nodes at each intersection; a diagonal steel bracing system works in tension and maintains rigidity. The floor is composed of prefabricated plastic-covered wood panels measuring 3.75 by 2.8 meters that are 60 millimeters thick, placed between the floor beams without screws. The ceiling is a system of clipped aluminum panels embedded with lights, an acoustical tile system, and water-filled copper pipes and coils that comprise part of the heating and cooling system.

Sobek worked with environmental engineer Matthias Schuler of Transsolar to coordinate the numerous energy-saving and sustainable building systems in this high-performance house. Cool-water and heat-storage systems were installed. Triple-glazed glass with inert gas and convection barriers keeps down heat transmission. The roof's flat solar-panel array, which supplies the energy for the mechanical ventilation system and the heat-pump system, maintains a constant temperature in the summer and winter and exchanges energy with the electric-grid-sensor technologies to power the lights and automatic doors.

"I don't know whether the generations coming after me will love the buildings or not," Sobek, says. "The only thing I want to give the buildings is the possibility that they can be taken away. If the people coming after us love them, the buildings are, of course, very much resilient, but resilient with technology. There is nothing that can degrade totally, but if it degrades, it can be replaced easily."

ABOVE: Street facade; staircase; intersection of rooms.

OPPOSITE: House illuminated with all of the structural elements visible.

JANE WERNICK ASSOCIATES

For Jane Wernick, engineering innovation starts with an initial conversation with the architects and the clients and with recognition of the architectural aspirations for a project. But the process does not end there. Wernick believes that projects benefit as the relationship with the architect develops and that the engineer can contribute to the overall design concept and, conversely, the architect begins to influence the engineer's ideas about structure. This level of collaboration and personal rapport is essential, and it has been an ingredient in Wernick's recipe for success in working with engineering colleagues such as Peter Rice, with whom she worked closely at Ove Arup & Partners, and with architectural firms such as Marks Barfield Architects, with whom, also while at Arup, she collaborated on the London Eye ferris wheel for the Millennium, a structural and industrial-design challenge.

Wernick began her career early, taking a year off from university to work as an intern at Arup—one of only eleven engineering firms in London that would take on pre-university women engineers at the time. She returned after graduation, working first on an unbuilt project called Kocommas, for the the King's Consultative Council in Saudi Arabia (1977–79). Wernick joined Frei Otto and Peter Rice on two domes, analyzing complex geometrical forms from first principles of basic geometries, not just relying on code. Joachim Schock, who had worked with Otto previously, led the buckling-stress analysis, and Wernick learned from him about complex shapes. She recognized in these early projects that "building codes don't cover funny shapes or nonstandard structures. If you move outside the normal structure, you have to think about analysis, loads, performance criteria, and stability." Deflections that are often ignored in calculations for "normal" buildings were necessary for these structures, which had comparatively larger deflections under similar loads and whose geometry causes them to behave in a different way. For nonlinear structures, the engineer's response and analysis also have to be nonlinear. The requirements of the project awakened Wernick's interest in innovative engineering and nonlinear methods.

After the Pompidou Center was completed, Peter Rice formed a small research-based firm, Piano & Rice, with Renzo Piano, the intention of which "was to examine

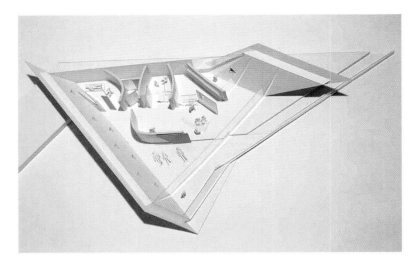

building and approach new projects with an engineering bias."[1] He also advised the Arup team that worked on the Kocommas project. In 1978 Wernick joined that team. Another of their commissions was a new car design for Fiat. Through an internal group at Fiat called IDEA, the team worked to develop an automobile chassis. One of the main structural requirements for a car frame is that it not be too flexible in torsion; the designers had to match the torsional strength of a standard car with their new frame concept. In the end, they developed ideas for the production process and incorporated separately painted components as well as proposals for alternative car-body shapes, presaging ideas of customized mass production in the automotive industry.

Wernick's interest in industrial production was sparked. But she pointed out the differences between engineering for industrial and architectural applications: "We could analyze the actual load case with the prototype because the manufacturer then puts the product into production. However, in a building you do a statistical analysis of the loads, and you see the live load on every floor, or on half of the floor, and then you add the wind load. A building is rarely going to be occupied to the maximum load, but because of code requirements everything is built for the most extreme load case, or worst-case scenario." Since full-scale building prototypes are rarely built, it is only possible to test loads mathematically and virtually. Out of these differences come lessons, however. For instance, engineers rarely get the chance to measure the deflections and stresses that the actual loads produce after a project is built, and Wernick suggests that they should, as is done in some seismically active regions, where accelerometers are used to monitor how buildings respond to the measured ground accelerations. Working with the Fiat team illuminated for Wernick, in a direct way, how "structural analysis is not a precise science, but difficult statistically; it is chaotic, and it is part craft."

At Arup, Wernick also worked with Peter Rice on the Grandes Serres of the Science and Technology Museum in La Villette, Paris (Adrien Fainsilber & Associés, 1986), as well as with Martin Francis and Ian Ritchie, who then established Rice, Francis, Ritchie (RFR) in Paris. Arup agreed to provide technical expertise and computing facilities on the project, because as Wernick recalls they did not have the computer software in Paris. Therefore, from London, she coordinated the computer analysis of the science center—three large glass boxes and a large fabric roof light supported by a tensegrity structure. The glass facade was a significant engineering innovation. The glass was used to carry the vertical loads of its own weight, and the wind loads were carried by cable trusses behind the glass. The glass was supported by glass bolts with a spherical bearing around each bolt. The weight is transferred through the bolt to a steel H-piece to another bolt in the plate above. Thus the glass iteslf carries the vertical loads. The spherical bearing also allows for rotational movement when the glass panels deflect under wind loading. This four-point support system became a trademark of Rice's glass curtain wall projects. Wernick completed most of the final computing on both the glass facades and the steel component of the roof structures and then worked in Paris for three months to prepare the final calculations to be submitted for official approval. This project expanded Wernick's perception of the potential for diverse structural effects to make a visual and spatial contribution to the architecture of a structure.

In the late 1970s and early 1980s, Arup would not send a woman to work in Africa or the Middle East, where they had many commissions, so Wernick went instead to Buffalo, New York, to work for Birdair, a manufacturer of tensile fabric structures. While there, she was able to further develop ideas about industrial production. In 1981 she went back to Arup, opening their Los Angeles office in 1986 and working collaboratively with architects on competition entries from the point of project conception. She also began teaching at the Southern California Institute of Architecture.

While in the United States, Wernick entered an ideas competition for a pedestrian bridge for the Grand Canyon sponsored by the Institution of Civil Engineers, called "An Image of the Bridge of the Future." Her project with the architects David Marks and Julia Barfield was a quadripedal bridge with a travelator, inspired by the body and bone structures of a four-legged animal, similar to those animal structures analyzed by D'Arcy Wentworth Thompson, a professor of zoology who combined disciplines of classics and math to analyze form. The prestressed bridge can be compared to Thompson's discussion of animal vertebrae in compression, and tendons that control the deformations under different loads.[2]

In 1989 Peter Rice suggested that she take a sabbatical from Arup, and she became a "Visiting Scholar" at Harvard's Graduate School of Design, taking courses in science, history of industry and technology, and on ideas of space. When Peter Rice became ill, she continued his projects while formulating her own method of collaborating with architects on their diverse design objectives.

She worked with Zaha Hadid on the Cardiff Opera House competition (1994) and the unbuilt KMR Art and Media Center competition for Düsseldorf (1989–93). For the project, Wernick developed a technique that systematized Hadid's irregular design. It was not essential for the columns to be on a rectilinear grid, but it would be easiest to build if the slabs they supported were generally flat. The thickness of the slab would depend on the maximum distance between columns. For a 25-centimeter-thick slab, the columns should not be more than 6 meters apart. For a 30-centimeter-thick slab, they should be no more than 7.5 meters apart. Otherwise additional beams would be added for support. This system allowed the engineers to work with the architects' holistic design without reanalyzing at every new iteration. Although some problems still required special solutions, her general system facilitated the design process and made the unusual buildings easier to build. Applying the same kind of ground rules and parameters, Wernick also designed the structure for Hadid's

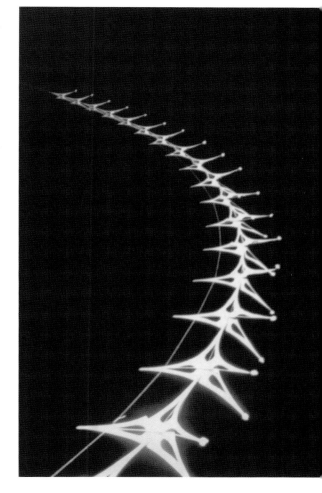

ABOVE: Pedestrian bridge designed to span the Grand Canyon, "An Image of the Bridge of the Future" competition, Marks Barfield Architects, 1989.

OPPOSITE: Maggie Center, Zaha Hadid, Fife, Scotland, 2007.

PAGE 208: London Eye, Arup, Marks Barfield Architects, London, 2000.

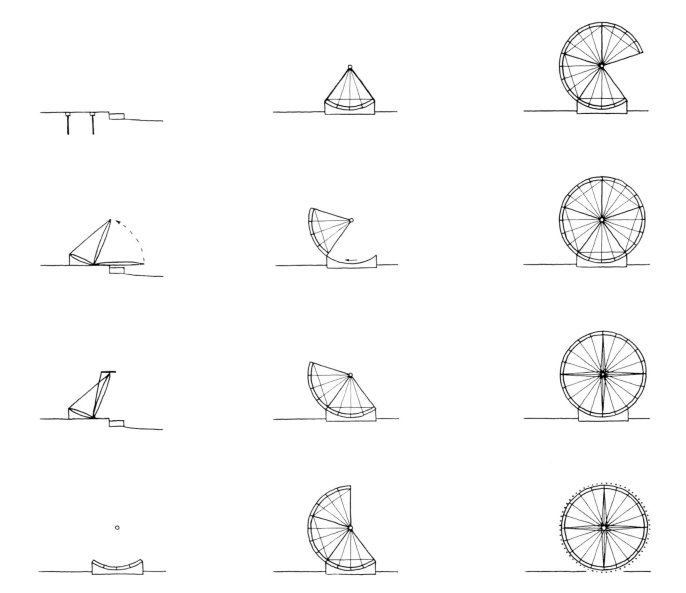

Maggie's Center in Fife, Scotland (2006), and an unbuilt scheme for the Holloway Road Bridge at the University of North London.

The most technically demanding and intellectually satisfying project Wernick designed at Arup was the Millennium Wheel in London, for which she was project director. Architects David Marks and Julia Barfield envisioned a 152-meter-high wheel. The challenge was to design a ferris wheel that was strong, did not buckle, and would be the tallest observation wheel built to date. Together they tried various strategies, and Wernick went to first principles of physics and structures, finding the bicycle wheel to be the most "economical and magical structure because it is a tensegrity structure with compression elements at the rim that never touch the spindle except from tension elements." On the wheel the main load came from the center of the spindle with 64 horizontal spokes maintaining the circular form to enable the vertical load to be carried. The wheel is supported on just one side. The main load comes from the rim in a structural open framework. The two pylons that support the wheel at the hub are placed on the land side of the wheel, with the spindle cantilevering out to allow the wheel to be suspended over the river. This also enhances the dynamic appearance of the structure. The original proposal was that five temporary radial trusses would be used to aid the construction of the wheel in segments, as shown in the construction sequence sketches. Arup's Advanced Technology group, which focused on car and product design, designed the thirty-two gondolas in the form of egg-shaped capsules with curved cast-glass windows, and they were attached to the outside of the rim in order to offer the best views from the wheel. In the end, some elements were simplified because of cost, but the Millennium Wheel has proved a permanent, successful structure on the London skyline.[3]

In 1998 Wernick left Arup to establish her own practice specializing in cultural and institutional projects. Her first projects were with architect

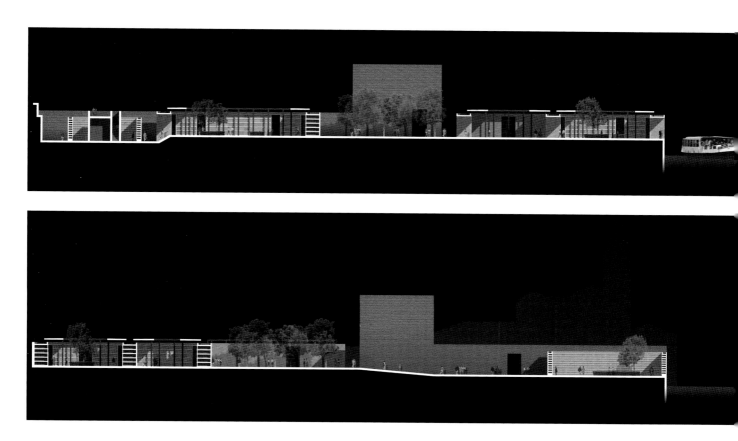

ABOVE: San Michele Island, David Chipperfield, Venice, Italy, 2006–8.

OPPOSITE: Diagrams of the London Eye, Arup, Marks Barfield Architects, London, 2000.

David Chipperfield, with whom she had collaborated while working at Arup, including on the extension of the cemetery on San Michele Island in Venice. Intensive structural and marine engineering helped to direct this infrastructure and landscape design project. The first phase, started in 2006, added tombs arranged into a series of volumes that express settlement and enclosure. The second phase will include the construction of a new island, with four new tomb buildings, which, unlike the original cemetery, will not be enclosed by a perimeter wall, but will lead to a garden that steps down to the water. Wernick continues to collaborate with Chipperfield on other projects as well.

Wernick sees the emergence of the design team "as a positive means of enriching design, and of resolving hugely complex design projects. This complexity is enmeshed in society, in a way which we may not have envisaged—some of it brought about by technological advances themselves." At times during projects when Wernick has found herself "agonizing about getting it right, when you know it won't work," she has reflected on the words of her mentor, Peter Rice, who would advise her to "let the architects draw it, but let them in on our secrets." Thus, her approach is to explain to the architects how the structure is working and why the structure is being used, helping the process past the point of difficulty.

Teaching, too, has changed the way Wernick communicates and works with architects, making the conversation more open. She took on a teaching appointment at the Architectural Association as a Diploma School Unit Master from 1998 to 2003. At special workshops she has conducted at universities in England, when she is on architectural juries, and in master classes she has taught, such as those at the Berlage Institute, in Rotterdam, she has conveyed the ideas that collaboration benefits the design and building process, that "the division between architecture and engineering" is an "arbitrary" one, and that the "broad band of shared knowledge" between the two professions enriches the process.

AERIAL WALKWAY, ROYAL BOTANIC GARDENS
KEW, LONDON

At Kew Gardens a temporary aerial walkway—constructed of scaffolding in 2003 and removed at the end of 2004—followed an elevated route through the Redwood Grove there. The Arboretum and Horticultural Services decided that a new, more permanent walkway designed by Marks Barfield Architects would allow visitors to experience the life of a tree canopy at a great height in a safe setting. After discussing the best route for the aerial walkway, the client decided to loop it through an area originally designed by Capability Brown (1716–1783). The area contains a wide variety of species, including a number of "Champion" (old growth) deciduous trees, and both near and far views of the plant life and habitat would vary seasonally.

Taking into consideration both the natural environment and the difficulties of maintaining a winding, narrow, elevated platform, the engineers and the architects strategized a simple, innovative design due to be completed in 2008. The structural constraints and the benefits of employing a series of standardized components prompted the use of a modular, kit-of-parts approach in the development of the walkway. This in turn created flexibility in the layout and an interesting, irregular geometry in which each node is the same but the sections connect at different angles. First Cor-Ten steel was selected as the main structural material. It forms a layer of oxidation on its surface that

ABOVE: Elevation.
BELOW: Site plan.
OPPOSITE: Rendering.

ABOVE: Renderings of the connecting platform lookouts, structural details, and Fibonacci series geometry.

OPPOSITE: Rendering of the pedestrian platform.

acts as a barrier to further corrosion—turning a reddish brown—and does not need maintenance. Suspension, mast, and deep-span structures were all discounted because masts and long spans would dominate the trees, and the walkway was intended to be subordinate to nature.

The Cor-Ten plates will be fabricated—large, tapering hollow triangular sections for the pylons, or hollow box sections, or thick plates. Piled foundations in reinforced concrete, about 600 millimeters in diameter—rather than wide footings, which would have disrupted a larger area of tree roots—anchored each of the pylons. In response to the issue of what the supports should be and how they should be connected, Wernick's team devised a system of structural pylons that supported the walkway at 15-meter intervals. Triangular, tapering pylons with secondary supporting branches were designed to carry the loads applied to the ring beams holding the viewing platforms at the top of each pylon.

The two horizontal trusses that support the deck and form the balustrades are clad with a steel mesh producing an arrangement of diagonals that was irregular but based on structural logic. The diagonals were not longer than 2.5 meters, and no more than three bracing elements met at one point. The deck itself works as a horizontal truss between the pylons that resists wind loads. The architects based the positions of the points along the trusses on the Fibonacci numerical series so that the ratio of each successive pair of numbers provides a proportion for the walkway module's geometry. Wernick drew on her experience with

Hadid's KMR project to establish rules for the geometry of the vertical trusses and determine the spacing of the elements. The design is such that there is repetition, but it looks quite random in its final assembly.

A main stair-elevator tower provides vertical circulation to the walkway. A central structural pylon with bifurcating, cantilevered supports carries the stairs and landings, thus providing unobstructed views in and out of the tower, as well as clear access at ground level. Two pylons with fixed guide rails carry the elevator and, in addition, add stability to the stair tower.

The close collaboration between Wernick and the architects resulted in an aerial walkway whose design was driven by the structural requirements and based on a geometric patterning.

FIGGE ART MUSEUM
DAVENPORT, IOWA

David Chipperfield Architects designed the Figge Art Museum on the banks of the Mississippi River in Davenport, Iowa, as an open and inviting museum—on a constrained urban site at the edge of a river that frequently floods. To meet the project's needs, Wernick developed a steel structural system that gave the building the potential to both be open and closed, recede and project. The building's aluminum-framed glass cladding—working in concert with the materials used on the building's interior skin—is variously transparent, translucent, and opaque, altering according to the function of the views from each facade. The building's internal circulation patterns are directed by the rectilinear main volume, which is pierced with a low, off-center tower. The effect—further enhanced by the ghostly appearance of the museum's sign between the inner and outer layers of glass—is one of weightlessness, belying the strength of the building's formidable structure.

Wernick resolved numerous structural challenges, including the raised basement, and the rigid grid specified by the architect. She worked through two schemes for two different sites, ultimately going with a shorter structure with a footprint of 315 by 120 feet above the first floor, and 15 feet wider at ground level. The site slopes up gently from the northern to the southern end and lies within a 100-year flood zone. The first floor is set above the 500-year flood zone. The building, organized on a 30-foot grid in the north–south direction and on a 21-foot grid in the east–west direction, is an adroit arrangement of structural beams, transfer trusses, and key supports.

The city's decision to allow the area to flood with river water determined the height of the building's river-facing main entrance, which was achieved by setting the building on a concrete plinth and creating a long, stepped approach. The parking garage was sited at a semi-basement level at the back and, in the front, in an open-sided concrete box. On the city side, the plinth becomes a wedge sloped to create a civic plaza and the parking area. The economical and restrained 100,000-square-foot space contains exhibition spaces in the two tower floors, which each have 24-foot-high ceilings and are reached by a large staircase. A light well, also used as a winter garden, provides views out to the river. The first floor contains galleries and event spaces; the ground floor includes visitor amenities, a café, and an orientation room. Bands of operable windows provide light and air into the office spaces, and a dark mesh screen between the inner skin and the outer skin of fritted-glass panes filter the sun as needed. The differences in glazing and coverings create a subtle shift of vision between outside and inside and between how the building is perceived during the day and night—with the smooth daytime facade morphing into a volumetric nighttime composition of recessed entrance spaces and glowing, lanternlike interiors.[4]

BELOW: Plan sketch showing the typical column and beam arrangement.

OPPOSITE: Street facade.

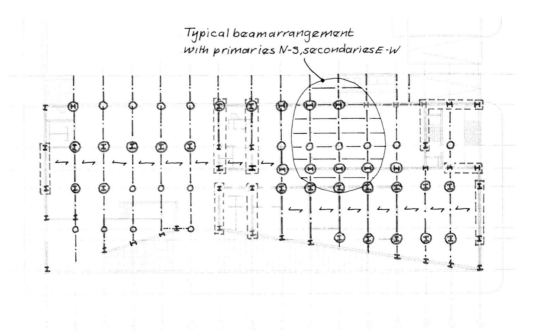

ABOVE: Steel beam structure in construction.

OPPOSITE: River facade illuminated.

Steel gave the team the freedom to utilize large transfer trusses three-quarters of the length of the building, allowing them to place a huge opening on the river side, for a staircase, and the facade setback. From the first floor up, a structural steel frame and metal decking is a composite slab. The primary beams span 30 feet and support the secondary beams, which are at 10-foot centers, whereas most of the secondary beams span 21 feet, opening up gallery spaces. The primary grid increases to accommodate the auditorium, and the rake of the auditorium is constructed using light steel framing that is supported by the primary structure. Computer models revealed that by using the transfer trusses, structural elements could be pushed and pulled, making the dramatic first-floor setback possible and allowing for the removal of regular columns on the grid in the ground-level parking garage.

The clear-span steel trusses used to form the roof of the permanent gallery on the second floor are 120 feet in length, and those for the roof of the temporary gallery and winter garden on the third floor are 63 feet in length. Lateral stability is provided by a series of braced bays in the east–west and the north–south directions, as well as between the third floor and the roof. The wind shears along this line are distributed among the bracing on some of the grid lines, by means of diaphragm action of the floor plate at the third floor. The lateral forces are carried by the bracing down to reinforced-concrete walls between the basement and the first floor and, from there, transferred to the ground by the foundation system.

Once key structural design decisions were made, Wernick monitored the design so that the structure worked correspondingly. This was accomplished using a feedback process between the architect and engineer collaborating in a way not unlike the constant push-pull of forces and materials.

NOTES

Unless otherwise indicated, quotations are from interviews the author conducted between 2005 and 2007.

INTRODUCTION

1. Arnold, Pacey, *Meaning in Technology* (Cambridge, Mass.: MIT Press, 1999), 13.

2. See for example, Sylvie Deswart and Bertrand Lemoine, in collaboration with Centre Georges Pompidou, *L'Architecture et les ingénieurs: deux siècles de construction* (Paris: Editions du Moniteur, 1979).

3. Ulrich Pfammatter, *The Making of the Modern Architect and Engineer* (Basel: Birkhäuser, 2000).

4. Tom F. Peters, *Transitions in Engineering: Guillaume Henri Dufour and the Early 19th Century Cable Suspension Bridges* (Basel: Birkhäuser, 1987).

5. Kenneth Frampton, *Studies in Tectonic Culture: The Poetics of Construction in Nineteenth and Twentieth Century Architecture* (Cambridge, Mass: MIT Press, 1995), 336.

6. See, for example, Nina Rappaport, "The Reception and Image of Modern Industrial Buildings." Docomomo International Conference Proceedings, Sept 2003: 239–43 and Reyner Banham, *Concrete Atlantis: U.S. Industrial Building and European Modern Architecture 1900–1925*. (Cambridge, Mass.: MIT Press, 1986).

7. Quoted in Tim and Charlotte Benton, eds., *Form and Function: A Sourcebook for the History of Architecture and Design*, 1890–1939 (London: Crosby Lockwood Staples, 1975), 33.

8. Le Corbusier, *Towards a New Architecture* (London: Architectural Press, 1927), 19 and 42.

9. Nina Rappaport, "Deep Decoration" in Emily Abruzzo and Jonathan Solomon, eds., *Decoration 306090*. (New York: Princeton Architectural Press, 2006), 95–103.

10. Manuel Delanda, discussion with author, New York, 2006.

11. Antoine Pevsner, Naum Gabo's brother, discussed the work in George Ricky, *Constructivism: Origins and Evolution* (New York: George Braziller, 1995), 30 and 31.

12. Jesse Reiser, *306090 Decoration*. Discussion at the Architectural League of New York, New York City, November 3, 2006.

13. Georgy Kepes, ed., *Structure in Art and in Science*, Vision + Value Series (New York: George Braziller, 1965).

14. James Bryan and Rolf Sauber, eds., "Interviews with Robert Le Ricolais" in *VIA, no. 2: Structures Implicit and Explicit*, (Graduate School of Fine Arts, University of Pennsylvania, Philadelphia,1973): 88.

15. Le Ricolais, "Interviews with Robert Le Ricolais," 88–91.

16. For example in, H. Kawamura, H. Ohmori, N. Kitu, "Truss topology optimization by a modified genetic algorithm," *Structural and Multidisciplinary Optimization*, 23 no. 6 (July 2002): 467–73.

ARUP

1. See Ove Arup, "The World of the Structural Engineer," Maitland Lecture presented at the Institute of Structural Engineers, London, 1968, published in *The Structural Engineer* (January 1969) and here cited from *The Arup Journal* 20, no. 1 (spring 1985).

2. Jack Zunz in Charles Jencks, "The Aesthetics of Engineering: Charles Jencks Interviews Jack Zunz," *Engineering and Architecture, Architectural Design* 57 (November–December 1987): 40.

3. Arup, "The World of the Structural Engineer," 2.

4. Arup, "The Key Speech." Lecture delivered at the Arup Organization, Winchester, England, July 9, 1970.

5. Arup, "The Key Speech."

6. Richard Hough, "Intuition in Engineering Design," in David Dunster, ed., *Arups on Engineering* (Berlin: Ernst & Sohn, 1997), 18–27.

7. Jencks, "The Aesthetics of Engineering," 42.

8. Brian Forster, "Lightweight Structures," in *Arups on Engineering*, 111–12.

9. Arup, "Sydney Opera House," *Architectural Design* (March 1965): 140.

10. Peter Rice, *An Engineer Imagines* (London: Ellipsis Books, 1996), 25–26.

11. Bryan Appleyard, *Richard Rogers: A Biography* (Boston: Faber and Faber, 1986), 211.

12. Robert Thorne, "Continuity and Invention," in *Arups on Engineering*, 234–61.

13. Arup, "The World of the Structural Engineer," 9.

14. Jencks, "The Aesthetics of Engineering," 45.

15. Cecil Balmond, discussion with author, New York, October, 2002.

16. A. Adão da Fonseca, A., Renato Bastos, A. Adão da Fonseca Jr., and Nuno Neves, "Design and Construction Method of the New Coimbra Footbridge" (paper presented at the Footbridge 2005: 2nd International Conference, Venice, Italy, December 6–8, 2005).

17. Rem Koolhaas, *Content: AMOMA* (Cologne: Taschen, 2004), 515.

18. Rory McGowan, "Structural Analysis" in *CCTV by OMA*, Special Issue of *A+U* (July 2005): 108.

19. Ruggero Lenci, "The Water Cube," *L'Arca* 187 (December 2003): 34–43.

20. Nina Rappaport, "Deep Decoration," 95–103.

21. Stuart Bull and Steve Downing, "Beijing Water Cube: The IT Challenge," *The Structural Engineer* (July 6, 2004): 23–26.

ATELIER ONE

1. Christopher Hight, "Dangerous Liaisons: The Art of Engineering after Truth and Beauty," *M'ARS: Magazine of the Museum of Modern Art Ljubljana* (February 2005): n.p.

2. Other engineers have researched ways to learn from structural failure, such as Henry Petroski, *Pushing the Limits: New Adventures in Engineering* (New York: Vintage Books, 2005), and Rowland Mainstone, *Developments of Structural Form* (Aldershot, England: Ashgate, 1999).

3. Roger Penrose, "Penrose Tilings of the Plane," (lecture presented at the MASS Colloquium, Pennsylvania State University, October 24, 2002), and Penrose, *Shadows of the Mind: A Search for the Missing Science of Consciousness* (Oxford: Oxford University Press, 1994).

4. Rappaport, "Deep Decoration," 98.

BOLLINGER + GROHMANN

1. André Chaszar, "Bernhard Franken," in Peter Cachola Schmal, ed., *Workflow: Architecture-Engineering, Klaus Bollinger + Manfred Grohmann* (Basel: Birkhäuser, 2004), 75.

2. Peter Cook, "Preface," in *Workflow,* 10.

3. Peter Cachola Schmal, "Coop Himmelb(l)au," in *Workflow,* 58.

4. Claudia Kugel, "Picture Palace," *Architectural Review* (July 1998): 54–58.

5. Schmal, "Coop Himmelb(l)au," in *Workflow,* 57.

6. Schmal, "Coop Himmelb(l)au," in *Workflow*, 60.

7. Le Ricolais, "Interviews with Robert Le Ricolais," 88–89.

8. Branko Kolarevic, ed., *Architecture in the Digital Age: Design and Manufacturing* (New York: Spon Press, 2003), 134.

9. Greg Lynn, FORM, project description, 2006.

10. Cook, "Preface," in *Workflow*, 10.

11. Cook, "Preface," in *Workflow*, 7.

12. Coop Himmelb(l)au, *Architecture is Now: Projects, (Un)buildings, Actions, Statements, Sketches, Commentaries, 1968–1983*, trans. Jo Steinbauer and Roswitha Prix (New York: Rizzoli, 1983), 198.

13. Cook, "Preface," in *Workflow*, 9.

14. Cook, "Preface," in *Workflow*, 9.

15. Ilka and Andreas Ruby, "Realities United," *Archplus* (October 2003): 10–15.

16. Peter Cook, Colin Fournier, and Klaus Kada, *Curves and Spikes*, exhibition catalogue, Aedes West Galerie, Berlin, March 7–April 20, 2003 (Berlin: Aedes, 2003).

BURO HAPPOLD

1. Edmund Happold quoted in "A Personal Perception of Engineering," *Architectural Design* 57, no. 11/12 (1987): 13.

2. Happold quoted in "A Personal Perception of Engineering," 19.

3. Happold quoted in "A Personal Perception of Engineering," 17.

4. Derek Walker, *Happold: The Confidence to Build* (Great Britain: Happold Trust Publications, 1997), 21.

5. Patrick Bellew emphasized this characteristic of Ted Happold in a discussion with the author. Bellew is a partner in the environmental engineering firm Atelier Ten, and studied with Happold at the University of Bath.

6. For more on Frei Otto, see Otto, *Tensile Structures: Design, Structure, and Calculation of Buildings of Cables, Nets, and Membranes*, vols. 1 and 2, trans. D. Ben-Yaakov and T. Pelz, (Cambridge, Mass: MIT Press, 1967, 1969, and 1973).

7. Frei Otto used soap film and bubbles to study the structures of the form.

8. Walker, *Happold*, 56–57.

9. Rice, *An Engineer Imagines*, 25. Happold is not credited for his work on the Pompidou Center project.

10. The institute has since merged with the engineering department at the University of Stuttgart and is now called the Institute for Lightweight Structures, Design and Construction (ILEK) under the direction of Werner Sobek.

11. Air-supported structures were a topic of research and investigation in general at the time in England. At Felix J. Samuely and Partners in 1971, Cedric Price with Frank Newby conducted "Air Structures," a survey commissioned by the Ministry of Public Building and Works for the Department of the Environment, which provides detailed technical information including history, applications, and costs.

12. Hooke's Law describes that stress is directly proportional to strain up to the limit of the proportionality constant. In an elastic body such as a spring, the deformation is proportional to the force applied.

13. Walker, *Happold*, 87.

14. Walker, *Happold*, 101.

15. Kenneth Powell, "Fruits of the Forest [Weald & Downland Open Air Museum]," *Architects' Journal* 216, no. 1 (July 4, 2002): 26–35.

16. Nicolas Pople, "Caught in the Web," *RIBA Journal* 108, no. 2 (February 2001): 39.

17. Buro Happold description in special issue on Foster's Great Court, *Architecture Today* no. 115 (February 2001): 32.

18. Chien Chu, "Engineering the Glass and Steel Roof to the British Museum Great Court Project, London, England, 2000," *Dialogue Structures* 32 (December 1999): 51–57.

CONZETT BRONZINI GARTMANN

1. Robert Maillart was a Swiss engineer much appreciated by modern artists and architects for his purity of form and efficiency of structure for infrastructural projects such as large span bridges in concrete. See Max Bill, *Robert Maillart* (Erlenbach-Zürich: Verlag für Architektur AG, 1949) and David P. Billington, *Robert Maillart's Bridges: The Art of Engineering*, (Princeton: Princeton University Press, 1979).

2. Critical regionalism, a term that both Kenneth Frampton and Alexander Tzonis use to describe the architect's design aesthetic based on local tradition and materials, can be applied to the engineering designs of Conzett Bronzini Gartmann.

3. See Jürg Conzett, "On Perception," in *Structure as Space: Engineering and Architecture in the Works of Jürg Conzett and His Partners*, ed. Mohsen Mostafavi (London: AA Publications, 2006), 256.

4. Nina Rappaport, "Working on the Railroad: A Swiss Train Station is Crafted on Site by Local Steelworkers," *Architecture* 92, no. 7 (July 2003): 56–61, and "The Swiss Section" exhibition, *Van Alen Report*, Van Alen Institute, New York, spring 2003.

5. Rappaport, "The Swiss Section," *Van Alen Report*.

6. Marcel Meili and Markus Peter, *Play Pentagon, Das new Fussballstadion auf dem Hardturm in Zürich*, (Zurich: Scheidegger & Spiess, 2005).

DEWHURST MACFARLANE AND PARTNERS

1. Gilles Deleuze and Felix Guattari, *A Thousand Plateaus: Capitalism and Schizophrenia* (Minneapolis: University of Minnesota Press, 1987).
2. Rice, *Engineer Imagines*, 107–13.
3. Annealing is a process in which the glass sheet is made at an even temperature and slowly cooled to harden. Tempered glass or toughened glass is even stronger than annealed and can be treated with a chemical component as well.
4. For additional information see www.pilkington.com.
5. Paul Finch, "Cool Quality," *The Architectural Review*, 218, no. 1304 (October 2005): 74–77.

EXPEDITION ENGINEERING

1. Arup, "The Key Speech."
2. See proceedings of "Footbridge 2005: 2nd International Conference, December 6–8, 2005, Venice, Italy." Two conferences have been held on footbridge designs that have special sessions on movement.
3. The Arup Journal, *Commerzbank, Frankfurt*, 32, no. 2, 1997: 3–11

LESLIE ROBERTSON

1. For more information on Fazlur Kahn, see Yasmin Sabina Kahn, *Engineering Architecture: The Vision of Fazlur R. Khan* (New York: W.W. Norton & Company, 2004).
2. Leslie Robertson, Gold Medal Address, published in *The Structural Engineer* (March 15, 2005): 21.
3. Leslie Robertson has written papers about airflow and building movement whereas the Danish engineers Jensen and Frank had previously developed the concept of the boundary-layer wind tunnel and they could replicate the pressures and suctions on small buildings, cited by Leslie Robertson, Gold Medal Address, 23.
4. Leslie Robertson, Gold Medal Address, 22.
5. For more information on the museum's design see, James A. Russell, "Pei Communes with Nature at Japan's Remote Miho Museum," *Architectural Record* 186, no. 8 (August 1998): 43.

GUY NORDENSON

1. Ezra Pound Cantos in Nadel, Ira B., The Cambridge Companion to Ezra Pound, (Cambridge: Cambridge University Press): 60.
2. Guy Nordenson, "With Great Joy and Expectations," brochure from Noguchi Fuller exhibition at the Noguchi Garden Museum, New York, 2006.
3. Developed by the French engineer Eugene Freyssinet, the system is a prestressed system of reinforcing bars for concrete construction.
4. Guy Nordenson, (lecture presented at the Architectural League, New York, July 20, 2006).
5. Guy Nordenson and Associates, *WTC Emergency: Damage Assessment of Buildings Structural Engineers Association of New York Inspection of September and October 2001*. Volume A Summary Report and Vols. B–F on DVD, SEAoNY, 2003.
6. Todd Gannon, *Steven Holl Architects/Simmons Hall*, (New York: Princeton Architectural Press, 2004).

RFR

1. Rice, *An Engineer Imagines*, 72.
2. The other directors include: Niccolò Baldassini, Henry Bardsley, Jean-François Blassel, Marc Chalaux, Benjamin Cimerman, Mitsu Edwards, Mathias Kutterer, Jean Le Lay, N'Dour N'Guissaly, Gilbert Plumet, Nicolas Prouvé, Bertrand Toussaint.
3. This fabrication process is described in detail in Peter Rice and Hugh Dutton, *Structural Glass* (London: E & FN Spon, 1993).
4. Rice, *An Engineer Imagines*, 76 and 77.
5. RFR, "La technique n'est pas neutre," *Le moniteur architecture, AMC* (September, 1999, n.100): 103–119.
6. The Footbridge 2002 conference in Paris brought together numerous engineers to discuss issues of movement in footbridge design.
7. Rice, *An Engineer Imagines*, 112.

SASAKI AND PARTNERS

1. Sasaki is applying principles of self-organization and evolution in nature and that of natural forms such as the spherical form of a soap bubble, whose surface tensions are minimized to ideas of structural efficiency and nonlinear structures similar to that of the ideas of Deleuze, or that of the structure of animals now possible through computation and genetic algorithms as new shape-design methods.
2. As noted in Mutsuro Sasaki, *Flux Structure: Musings on Shape Design* (Tokyo: TOTO Ltd., 2005).
3. "Secret Structure, Dialogue between Sasaki, Sejima, and Nishizawa," in *Hunch: The Berlage Institute Report*, no. 1 (1999): 147.
4. Mutsuro Sasaki, interview by Marc Guberman from questions written by the author, Tokyo, August 2006.
5. Sasaki, *Flux Structure*, 11.
6. Sasaki, interview, August 2006.
7. "Secret Structure, Dialogue between Sasaki, Sejima, and Nishizawa," 147.
8. "Secret Structure, Dialogue between Sasaki, Sejima, and Nishizawa," 149.
9. Sasaki, *Flux Structure*, 21.
10. Sasaki interview, August 2006.
11. Sanford Kwinter, "The Sendai Solid," in *Case: Toyo Ito, Sendai Mediatheque*, eds. Ron Witte with Hiroto Kobayashi, (Munich Prestel Verlag, Harvard University Graduate School of Design, 2002), 32.
12. Sasaki, "Structural Design for the Sendai Mediateque," in Kwinter, 43.
13. Sasaki interview, August 2006.
14. Sasaki interview, August 2006.
15. Sasaki, *Flux Structure*, 73.
16. Sasaki, *Flux Structure*, 83.
17. Sasaki, I Project, *A+U*, 404, no. 4 (May 2004): 26–43.
18. Sasaki, *Flux Structure*, 63.
19. Sasaki, "I Project," *A+U*, 404, no. 4 (May 2004): 26-43.
20. Manfred Bollinger, discussion with author, New York, November 2006.

SCHLAICH BERGERMANN UND PARTNER

1. See Jörg Schlaich and Rudolph Bergermann. *Leicht Weit/Light Structures*, exhibition catalogue, Deutches Architektur Museum, Frankfurt am Main, November 22, 2003–February 8, 2004 (Munich: Prestel, 2003).
2. Institute for Lightweight Structures, Design and Construction (ILEK).
3. Schlaich and Bergermann, 112-14.
4. Lecture at the Yale School of Architecture, New Haven, Connecticut, November 15, 2004. As Schlaich presented this project he noted that the design was perhaps not justified, and over done, although his firm can engineer numerous forms that are out of the norm.
5. Herbert Klimke, "Neue Messe Mailand—Netzstruktur und Tragverhalten einer Freiformflache," *Stahlbau* 73, Heft 8, 2004.
6. Schlaich and Bergermann. *Leicht Weit/Light Structures*, 246–47.
7. Hans Schober, "The Berlin Connection" *Civil Engineering* (August 2006): 43–49, 81.

WERNER SOBEK

1. An exhibition, *Archi-Neering*, Helmut Jahn and Werner Sobek, was held at the Stadtisches Museum Leverkusen Schloss Morsbroich, Germany, June 6–September 12, 1999. A catalogue of the same name was published by Hatje Cantz Verlag, 1999.
2. *Archi-Neering*, Helmut Jahn and Werner Sobek, 92.
3. "Bauen in der Zukunft," *Detail* 41, no. 8 (December 2001): 1454-1457.
4. Conway Lloyd Morgan, *Show Me The Future: Engineering and Design by Werner Sobek* (Ludwigsburg, Germany: Avedition, 2004), 138.
5. *UN Studio: Mercedes-Benz Museum: Design Evolution*, exhibition catalogue, Wechsel Raum, Bund Deutscher Architekten, BDA, December 18, 2005 to January 23, 2006 (Ludwigsburg, Germany: Avedition, 2006).
6. Ben van Berkel, discussion with the author, January 2007.

JANE WERNICK

1. For more information on this design see Rice, *An Engineer Imagines*, 135.
2. D'Arcy Wentworth Thompson, *On Growth and Form* (Cambridge: Cambridge University Press, 1961), 250.
3. For more information on the Eye and team involved in its design, fabrication, and construction, see Jane Wernick, *Architecture Today* 108: 44–47 and Sutherland Lyall, *Masters of Structure*, (London: Laurence King Publishing Ltd. 2002),162–71.
4. Suzanne Stephens, "David Chipperfield's luminous glass structure brings a clarity and rigor to the new Figge Art Museum in Davenport, Iowa," *Architectural Record* 193 (November 2005): 116–21.

BIBLIOGRAPHY

ARUP

Appleyard, Bryan. *Richard Rogers: A Biography.* Boston: Faber and Faber, 1986.

The Arup Journal 35, no. 2 (February 2000). Millennium issue.

Arup, Ove N. "Art and Architecture, The Architect-Engineer Relationship." Lecture presented at RIBA, on receiving the Royal Gold Medal, London, June 21, 1966. Published in *RIBA Journal* (August 1966): 350–59.

_____. "The Engineer and the Architect." *The Architect's Yearbook* 9 (1960): 167–72.

_____. "The Key Speech." Lecture delivered at the Arup Organization, Winchester, England, July 9, 1970.

_____. "Shell Construction." *Architectural Design* 17, no. 11 (1947).

_____. "The World of the Structural Engineer." Maitland Lecture presented at the Institution of Structural Engineers, London, 1968. In *The Structural Engineer*. London: Institution of Structural Engineers, January 1969.

Dunster, David, Ed. *Arups On Engineering.* Berlin: Ernst & Sohn, 1997.

"La Foresta nella rete: Hong Kong Park Aviary." *L'Arca* 57 (February 1992): 82–85.

Jencks, Charles. "The Aesthetics of Engineering: Charles Jencks Interviews Jack Zunz." *Architectural Design* 57 (November-December 1987): 36–48.

Jones, Peter. *Ove Arup: Masterbuilder of the Twentieth Century*. New Haven: Yale University Press, 2006.

Koolhaas, Rem. *Content: Triumph of Realization*. Cologne: Taschen, 2004.

CECIL BALMOND

Balmond, Cecil. *Informal*. Munich: Prestel, 2002.

"CCTV by OMA." *A+U: Architecture and Urbanism.* Special issue (July 2005).

"Conversation: Cecil Balmond and Toyo Ito, 'Concerning Fluid Spaces.'" *A+U: Architecture and Urbanism* 404 (May 2004): 44–53.

"OMA: CCTV Television Station and Headquarters, Beijing, China, 2002–08." *Lotus International* 123 (2004): 96–99.

Rappaport, Nina. "Cecil Balmond." *Schweizer Ingenieur und Architekt*, 35 (September 3, 1999): 9–12.

_____. "Dialogue between Cecil Balmond and Manuel Delanda." *Constructs*, Yale School of Architecture (Spring 2003): 14–15.

_____. "H_Edge." *Architect's Newspaper*, New York edition (October 16, 2006): 17.

_____. Interview with Cecil Balmond. *Constructs*, Yale School of Architecture (Fall 1999): 3.

_____. Interview with Cecil Balmond. *Metropolis Magazine* (March 2003): 109–12.

TRISTRAM CARFRAE

Bull, Stewart, and Steve Downing. "Beijing Water Cube: The IT Challenge." *The Structural Engineer* 82, no. 13 (July 6, 2004): 23–26.

Lenci, Ruggero. "The Water Cube." *L'Arca* 187 (December 2003): 34–43.

Woolnough, Paul. "Swimming in Bubbles." *Engineers Australia* (September 2004): 34–38.

ATELIER ONE

Bates, Donald L., and Peter Davidson. "Federation Square Melbourne Australia, Lab Architecture Studio." *Assemblage* 40 (December 1999): 58–67.

Dawson, Susan. "All the World's a Stage." *Architects' Journal* 206, no. 2 (July 10, 1997): 37–39.

Davey, Peter. "Lab Experiments, Urban Regeneration, Melbourne, Australia." *Architectural Review Australia* 213, no. 1275 (May 2003): 56–63.

Hight, Christopher. "Dangerous Liaisons: The Art of Engineering after Truth and Beauty." *M'ARS: Magazine of the Museum of Modern Art Ljubljana* (February 2005).

Macneil, James. "Concert Pitch: Eisteddfod Pavilion, Llangollen." *Building* 257, no. 9 (February 28, 1992): 36–39.

Melvin, Jeremy. "Federation Square, Melbourne." *Architectural Design* 73, no. 2 (March-April 2003): 103–9.

Penrose, Roger. "Penrose Tilings of the Plane." Lecture presented at the MASS Colloquium, Pennsylvania State University, October 24, 2002.

_____. *Shadows of the Mind: A Search for the Missing Science of Consciousness*. Oxford: Oxford University Press, 1994.

Slessor, Catherine, "Crimson Vortex." *Architectural Review* 206, no. 1231 (September 1999): 38–39.

Stokdyk, John. "Talent Spot: Neil Thomas." *Building* 257, no. 26 (June 26, 1992): 42–43.

BOLLINGER + GROHMANN

Bollinger, Klaus. "Rollenspiel als Realitatseinsteig." *Architektur & Bau* Forum 4 (1997): 81–84.

Bollinger, Klaus, Bernd Heidlindemann, Harald Kloft, Matthias Michel, and Matthias Stracke. "Turning Free Forms into Feasible Structures Structural Engineering in Digital Workflow." In *Architecture as Brand Communication*. Basel: Birkhäuser, 2002.

Cook, Peter, Colin Fournier, and Klaus Kada. *Curves and Spikes*. Exhibition catalog, Aedes West Galerie, Berlin, March 7-April 20, 2003. Berlin: Aedes, 2003.

"Coop Himmelb(l)au: Die Dynamik des Zwischen-Raumes." *Architektur Aktuell*, 6/7 (1998): 38–49.

"Coop Himmelb(l)au: UFA-Palast 'Kristall' in Dresden." *Baumeister* 95, no. 8 (August 1998): 46–54, 67.

Jaeger, Falk. "UFA-Kinozentrum, Dresden." *Bauwelt* 88, no. 4 (January 1997): 166–67.

Jones, Peter Blundell. "Alien Encounter: Art Museum, Graz, Austria." *Architectural Review* 215, no. 1285 (March 2004): 44–53.

Kugel, Claudia. "Picture Palace." *Architectural Review* 204, no. 1217 (July 1998): 54–58.

"Novartis Pharma Headquarters, Basel 2002." *El Croquis* 121/122 (2004): 168–73.

Schmal, Peter Cachola. "Archigram Goes Bilbao: Bollinger & Grohmann als Partner bei der Realisierung." *Archithese* 32, no. 6 (November-December 2002): 42–47.

_____, ed. *Workflow: Architecture-Engineering, Klaus Bollinger + Manfred Grohmann*. Basel: Birkhäuser, 2004.

"Sejima + Ryue Nishizawa/SANAA: Novartis Campus." *GA Document* 79 (May 2004): 96–97.

BURO HAPPOLD

Castle, Helen. "The Buro Happold Tapes." *Architectural Design* 72, no. 5 (September-October 2002): 67–78.

Chu, Chien. "Engineering the Glass and Steel Roof to the British Museum Great Court Project, London, England, 2000." *Dialogue Structures* 32 (December 1999): 51–57.

Davey, Peter. "In the Public Eye: Underground Station, Stuttgart, Germany." *Architectural Review* 213, no. 1274 (April 2003): 66-69.

Dixon, Jeremy. "Foster's Great Court." *Architecture Today* 115 (February 2001): 14–36.

Gill, Colin. "Tensyl di Buro Happold: An Interactive Graphic System." *L'Arca* 73 (July-August 1993): 68–73.

Happold, Edmund. "A Personal Perception of Engineering." *Architectural Design* 57, no. 11/12 (1987).

Hart, Sara. "A Brilliant Shell Game at the British Museum." *Architectural Record* 189, no. 3 (March 2001): 149–54.

Hart, Sara. "Foster & Partners Revives that Imperial Dowager, the British Museum, for Life in the 21st Century, While Restoring its Lost Architectural Past." *Architectural Record* 189, no. 3 (March 2001): 114–21.

"Holding Court." *Architectural Review* 209, no. 1247 (January 2001): 15.

Johnson, Steve. "The Architecture Ensemble." *Architectural Design* 75, no. 4 (July-August 2005): 96–99.

Littlefield, David. "Twisted Ideas." *Blueprint* 215 (January 2004): 44–47, 49–50.

Melvin, Jeremy. "Downland Gridshell at the Weald and Downland Open Air Museum, Chichester." *Architectural Design* 73, no. 1 (January-February 2003): 106–10.

Pople, Nic. "Off the Grid: Weald and Downland Gridshell." *RIBA Journal* 109, no. 5 (May 2002): 36–44.

Powell, Kenneth. "Fruits of the Forest." *Architects' Journal* 216, no. 1 (July 4, 2002): 26–35.

Walker, Derek, and Bill Addis. *Happold: The Confidence to Build*. London: Happold Trust Publications, 1997.

Weinstock, Michael. "Engineering Exigesis: Soft Materials, Strong Structures." *Architectural Design* 72, no. 2 (March 2002): 119–24.

CONZETT BRONZINI GARTMANN

Bill, Max. *Robert Maillart*. Erlenbach-Zürich: Verlag für Architektur AG, 1949.

Conzett, Jürg. "Jürg Conzett & Partners: Two Bridges." *AA Files* 41 (Summer 2000): 3–8.

Deplazes, Andrea, ed. *Constructing Architecture: Materials, Processes, Structures: A Handbook*. Basel: Birkhäuser, 2005.

Meili, Marcel, and Peter Markus. *Play Pentagon, Das neu Fussballstadion auf dem Hardturm in Zürich*. Zürich: Scheidegger & Spiess, 2005.

Miller, Quintus. "Le projet passe pas la critique." *Matières* 6 (2003): 16–17.

Mostafavi, Mohsen, ed. *Structure as Space: Engineering and Architecture in the Works of Jürg Conzett and His Partners*. London: AA Publications, 2006.

Pages, Yves. "Entretien avec Jürg Conzett: 'J'admets comme seul principe valable celui de la construction sensée.'" *AMC, Le moniteur architecture* 110 (October 2000): 76–77.

Rappaport, Nina. "Suransuns Footbridge: Viamala, Switzerland." *Architecture* 89, no. 10 (October 2000): 108–11.

_____. "Working on the Railroad: A Swiss Train Station is Crafted on Site by Local Steelworkers." *Architecture* 92, no. 7 (July 2003): 56–61.

Wagner, George. *Barkow Leibinger: Werkbericht 1993–2001*. Basel: Birkhäuser, 2001.

DEWHURST MACFARLANE

Dawson, Susan. "Zen and the Art of Glazing." *AJ Focus* 5, no. 4 (April 1991): 21–25.

Dewhurst Macfarlane and Partners, Jonathan Sakula, and Philip Wilson. "Structural Glass Engineering." In *Glass: Structure and Technology in Architecture*. Munich: Prestel, 1998.

Finch, Paul. "Cool Quality." *Architectural Review* 218, no. 1304 (October 2005): 74–77.

Maxwell, Robert. *Rick Mather Architects*. London: Black Dog Publishing, 2006.

O'Callaghan, James. "Structural Glass Facade for the Samsung Jong-Ro Building, Seoul, Korea, 1999." *Dialogue* (December 1999): 66–74.

EXPEDITION ENGINEERING

Booth, Robert. "Wise Up to Risks." *Architects' Journal* 212, no. 5 (August 3–10, 2000): 24–25.

Richardson, Vicky. "Triple Jump." *RIBA Journal* 106, no. 9 (September 1999): 18–19.

Wise, Chris. "The American Air Museum, Duxford." *The Arup Journal* 32, no. 3 (March 1997): 10–16.

_____. "Commerzbank, Frankfurt." *The Arup Journal* 32, no. 2 (February 1997): 2–10.

_____. "Drunk in an Orgy of Technology." *Architectural Design* 74, no. 3 (May-June 2004): 54–63.

_____. "The Man Who Fell to Earth." *RIBA Journal* 107, no. 1 (January 2000): 52–56.

_____. "Torre de Collserola, Barcelona." *The Arup Journal* 27, no. 1 (Spring 1992): 3–7.

GUY NORDENSON AND ASSOCIATES

Holl, Steven. "Experiments in Porosity." Martell Lecture presented at the School of Architecture and Planning, University at Buffalo, the State University of New York, April 1, 2005.

Jones, Wes. "Three Engineers (Sitting Around Talking)." *ANY: Architecture New York* 10 (1995): 50–55.

Lyall, Sutherland. "Using the Power of Design Prefabrication." *Architects' Journal* 219, no. 9 (March 4, 2004): 4–6.

Nordenson, Guy. "City Square: Structural Engineering, Democracy, and Architecture." *Grey Room* 7 (Spring 2002): 103–6.

_____. "Four Experimental Projects." *Dialogue Structures* (1999): 58–64.

_____. "Notes on Light and Structure." *Architectural Design* 67, no. 1/2 (1997): 2–14.

Popham, Peter. "Richard Meier Achieves a Baroque Sense of Space and Light with Concrete Construction in the Jubilee Church in Rome." *Architectural Record* 192, no. 2 (February 2004): 101–6.

Richards, Ivor. "Instrument of Light." *Architectural Review* 1286 (April 2004): 48–53.

LESLIE E. ROBERTSON ASSOCIATES

Faschan, William J., and Daniel A. Sesil. "Rising High." *Urban Land* (November-December, 2000): 62–67, 133.

"Is The Skyscraper Doomed? Symposium at National Building Museum Explores the Question." *Inform* 12, no. 4 (2001): 8–9.

Merkel, Jayne. "The New Age of the Engineer." *Oculus* 62, no. 7 (March, 2000): 8–11.

Robertson, Leslie E. "The Miho Museum Bridge, Shiga-raki, Japan." *Structure* (Winter 1998): 19–23.

Robertson, Leslie E. "Rising to the Challenge." *The Structural Engineer* 83, no. 6 (March 15, 2005): 20–27.

Russell, James A. "Pei Communes with Nature at Japan's Remote Miho Museum." *Architectural Record* 186, no. 8 (August 1998): 43.

Sesil, Daniel A., and Onur Güleç. "Commanding Presence." *Civil Engineering* 75, no. 3 (March 2005): 43–49.

Wilson, Forrest. "A Star in the East." *Blueprints* 9, no. 4 (Fall 1991): 1–6.

RFR

Baldassini, Niccolò. "Bridges of Glass." *Detail* 39, no. 8 (December 1999): 1428–32.

———. "La Genialità di Peter Rice: Aesthetics in Engineering." *L'Arca* 76 (November 1993): 46–57.

"Bardsley, Directeur du Bureau d'Études RFR." *AMC, Le moniteur architecture* 71 (May, 1996): 73.

Blassel, Jean-François. "Innovation et Performance: Intelligent Building Design." *AMC, Le moniteur architecture* 130 (January 2003): 26.

Davoine, Gilles. "Paul Andreu, Aérogare 2F à Roissy." *AMC, Le moniteur architecture* 87 (March 1998): 46–57.

Desvigne, Michel, and Michel Virlogeux. "Double viaduc des angles, Avignon." *L'Architecture d'aujourd'hui* 335 (July-August 2001): 68–73.

"Details: Passerelle de Bercy-Tolbiac sur la Seine à Paris." *AMC, Le moniteur architecture* 132 (March 2003): 98–99.

"The Grand Louvre Inverted Pyramid, Paris, France, 1993." *A+U: Architecture and Urbanism* 327, no. 12 (December 1997): 98–105.

"Huit portraits d'ingénieurs." *Architecture intérieure-créé* 277 (1997): 30–35.

Kronenburg, Robert. "RFR." *Architectural Design* 65 (September-October 1995): 9–10.

Ménard, Jean-Pierre. "Les façades double peau." *AMC, Le moniteur architecture* 101 (October 1999): 76–92.

Picon, Antoine. "RFR: Techniques singulières." *AMC, Le moniteur architecture* 100 (September 1999):103–19.

Rappaport, Nina. "Engineering Intuition: RFR." *Schweizer Ingenieur und Architekt* 23 (June 9, 2000): 8–14.

Rice, Peter. *An Engineer Imagines*. London: Ellipsis, 1996.

Rocca, Alessandro. "Peter Rice, Poet of Brutalism." *Lotus* 78 (1993): 6–39.

"TGV Station, Lille, France 1994." *A+U: Architecture and Urbanism* 327, no. 12 (December 1997): 110–17.

Vaudeville, Bernard. "Le rôle éminent du calcul." *AMC, Le moniteur architecture* 87 (March 1998): 58–61.

Vitta, Maurizio. "Presenza e apparenza: Birdhouse Project Space #5." *L'Arca* 134 (February 1999): 4–15.

SASAKI AND PARTNERS

"EPFL Learning Center/SANAA." *GA Japan: Environmental Design* 73 (March-April 2005): 14–21.

Hasegawa, Yuko. *Kazuyo Sejima + Ryue Nishizawa/SANAA*. Milan: Electa, 2005.

"Huit portraits d'ingénieurs." *Architecture intérieure-créé* 277 (1997): 30–35.

"I-Project, Fukuoka, Japan: Toyo Ito Architect & Associates." *GA Document* 79 (May 2004): 44–49.

"I-Project in Fukuoka, 2002–2005." *El Croquis* 123, no. 5 (2005): 294–309.

Kondo, Tetsuo. "EPFL Learning Center, Lausanne, Switzerland: Kazuyo Sejima + Ryue Nishizawa/SANAA." *GA Document* 85 (May 2005): 34–39.

"Mutsuro Sasaki, Vision of Structure." *Kenchiku Bunka* 54, no. 632 (June 1999).

Sasaki, Mutsuro. *Flux Structure: Musing on Shape Design*. Tokyo: Toto, 2005.

———. "I-Project: Fukuoka, Japan, 2002–2005." *A+U: Architecture and Urbanism* 404, no. 5 (May 2004): 26–43.

"Secret Structure: Dialogue Between Sasaki, Sejima, and Nishizawa." *Hunch: The Berlage Institute Report* 1 (1999): 144–51.

"Special Feature: Mutsuro Sasaki, Vision of Structure: Dialogues with Kazuyo Sejima, Ryue Nishizawa, Yutaka Saito, Kazuhiko Namba, and Toyo Ito." *Kenchiku Bunka* 54, no. 632 (June 1999): 15, 17–20.

Witte, Ron, and Hiroto Kobayashi, eds. *Case: Toyo Ito, Sendai Mediatheque*. Munich: Prestel with Harvard University Graduate School of Design, 2002.

SCHLAICH BERGERMANN UND PARTNER

"Couverture temporaire des arènes romaines, Nîmes." *L'Architecture d'aujourd'hui* 327 (April 2000): 28–29.

Geipel, Kaye. "2006 München: Zum Entscheid für das Stadion." *Bauwelt* 93, no. 11 (March 15, 2002): 20–21.

Jaeger, Falk. "Die Fussgängerbrücken: Volkwin Marg, von Gerkan Marg und Partner, und Jörg Schlaich, Schlaich, Bergermann und Partner." *Baumeister* 97, no. 7 (July 2000): 49–51.

"Jörg Schlaich, Schlaich Bergermann und Partner: Torre panoramica nel Parco Killesberg, Stoccarda, Germania 2001." *Casabella* 67, no. 711 (May 2003): 80–83.

"Light Structures: Tensile, Space, Pneumatic Structures." *Zodiac* 21 (1972).

Rappaport, Nina. *Light Structures: The Work of Jörg Schlaich & Rudolf Bergermann*. Exhibition catalog, Yale School of Architecture Gallery, November 15, 2004–February 4, 2005. New Haven: Yale School of Architecture, 2004.

Schlaich, Jörg. "Die Einheit von Form und Konstruktion: Formentwicklung im Leichtbau." *Arch Plus* 159/160 (May 2002): 26–33.

———. "A Plea for Concrete Construction in Keeping with the Nature of the Material." *Detail* 41, no. 1 (January-February 2001): 28–29.

———. "Seilbrücken: Zur konstruktiven Durchbildung von Seilkonstruktionen." *Deutsche Bauzeitung* 126, no. 4 (April 1992): 72–77, 80–85.

Schlaich, Jörg, and Rudolph Bergermann. *Leicht Weit/Light Structures*. Exhibition catalog, Deutches Architektur Museum, Frankfurt am Main, November 22, 2003-February 8, 2004. Munich: Prestel, 2003.

Schober, Hans, Kai Kurschner, and Hauke Jungjohann. "Neue Messe Mailand-Netzstruktur und Tragverhalten einer Freiformflache." *Stahlbau* 73, no. 8 (2004): 541–52.

Weischede, Dietger. "Zusammenarbeit Architect-Ingenieur." *Architekt* 11 (November 1996): 693–96.

WERNER SOBEK

"Archi e volte: New Glass Techniques." *L'Arca* 139 (July-August 1999): 76–79.

Baus, Ursula. "La Maison Sobek, Stuttgart, Allemagne." *L'Architecture d'aujourd'hui* 339 (March-April 2002): 32–34.

Baus, Ursula, and Werner Sobek. "Leicht zu Bauen ist nicht leicht: zur Charakteristik von Membran und Seilnetzkonstruktionen." *Deutsche Bauzeitung* 127, no. 9 (September 1993): 18–23.

Blaser, Werner. *R128 by Werner Sobek*. Basel: Birkhäuser, 2001.

"Exhibition Pavilion in Frankfurt: Sobek & Rieger." *Detail* 36, no. 8 (December 1996): 1240–44.

Jaeger, Falk. "Enkel im Geist Buckminster Fullers: Der Ingenieur-Architekt Werner Sobek." *Archithese* 32, no. 6 (November-December 2002): 48–51.

Jahn, Helmut, and Matthias Schule. *Post Tower*. Basel: Birkhäuser, 2004.

Jahn, Helmut, and Werner Sobek. *Archi-neering*. Ostfildern, Germany: Hatje Cantz, 1999.

Kuhnert, Nikolaus, and Angelika Schnell. "Sobek's Sensor." *Arch Plus* 157 (September 2001): 24–29.

Morgan, Conway Lloyd. *Show Me the Future: Engineering and Design by Werner Sobek*. Ludwigsburg, Germany: Avedition, 2004.

"Removable Roof Covering to the Arena in Nîmes." *Detail* 34, no. 6 (December 1994-January 1995): 819–24.

Sobek, Werner. "Building in the Future." *Detail* 41, no. 8 (December 2001): 1454–57.

_____. "Pneu und Schale: Betonschalen und pneumatish vorgespannte Membranen." *Deutsche Bauzeitung* 124, no. 7 (July 1990): 66–71.

_____. "Technological Principles of Membrane Structures." *Detail* 34, no. 6 (December 1994-January 1995): 776–80.

Sobek, Werner, and Anja Eckert. "Trapezkunst und Wellenspiel: Blech als tragender Baustoff." *Deutsche Bauzeitung* 129, no. 1 (January 1995): 120–27.

Sobek, Werner, and Martin Speth. "Von der Faser zum Gewebe: Textile Werkstoffe im Bauwesen." *Deutsche Bauzeitung* 127, no. 9 (September 1993): 74–81.

Sobek, Werner, and Mathias Kutterer. "Anders Bauen mit Glas." *Arch Plus* 144/145 (December 1998): 24–27.

UN Studio: Mercedes-Benz Museum: Design Evolution. Exhibition catalog, Wechsel Raum, Bund Deutscher Architekten, BDA, December 18, 2005-January 23, 2006. Ludwigsburg, Germany: Avedition, 2006.

JANE WERNICK ASSOCIATES

Dixon, Jeremy. "Full Circle: Marks Barfield on the South Bank." *Architecture Today* 108 (May 2000): 44–47.

Slavid, Ruth. "A Flowering Career." *Architects' Journal* 203, no. 5 (January-February 1996): 22–23.

Smith, Dennis B., and Paul Warner. "Practicing a Multidisciplinary Architecture: A Discussion with John Pringle and Jane Wernick." *Dimensions* 10 (1996): 75–79.

Stephens, Suzanne. "David Chipperfield's Luminous Glass Structure Brings a Clarity and Rigor to the New Figge Art Museum in Davenport, Iowa." *Architecural Record* 193, no. 11 (November 2005): 116–21.

Wernick, Jane. "Technology as Catalyst." In *Arups On Engineering*. Berlin: Ernst & Sohn, 1997.

Wernick, Jane, and Sarah Wigglesworth. "Clearwater Garden: Design Research and Collaboration." *ARQ: Architectural Research Quarterly* 6, no. 3 (September 2002): 214–29.

GENERAL

"Architects and Engineers." *Detail* 45, no. 12 (2005). Special issue.

"Architekt und Ingenieur." *Archithese* 32, no. 6 (November-December 2002).

"Balthasar-Newmann-Preis 2002." *Deutsche Bauzeitung* 136, no. 6 (June 2002): 43–108.

Banham, Reyner. *The Architecture of the Well-Tempered Environment*. 2nd ed. Chicago: University of Chicago Press, 1984.

_____. *Concrete Atlantis: U.S. Industrial Building and European Modern Architecture 1900–1925*. Cambridge, Mass.: MIT Press, 1986.

Benton, Tim and Charlotte, eds. *Form and Function: A Sourcebook for the History of Architecture and Design, 1890–1939*. London: Crosby Lockwood Staples, 1975.

Benyus, Janine M. *Biomimicry: Innovation Inspired by Nature*. New York: Perrenial, 2002.

Billington, David P. *Robert Maillart's Bridges: The Art of Engineering*. Princeton: Princeton University Press, 1979.

Blaser, Werner, ed. *Santiago Calatrava: Engineering Architecture*. Rev. ed. Basel: Birkhäuser, 1990.

Bridging the Gap: Rethinking the Relationship of Architect and Engineer. Proceedings of the Building Arts Forum/New York, Guggenheim Museum, New York, April, 1989. New York: Van Nostrand Reinhold, 1991.

Bryan, James, and Rolf Sauber, eds. "Interviews with Robert Le Ricolais." In *VIA 2: Structures Implicit and Explicit*. Philadelphia: University of Pennsylvania Graduate School of Fine Arts, 1973.

Deleuze, Gilles, and Felix Guattari. *A Thousand Plateaus: Capitalism and Schizophrenia*. Minneapolis: University of Minnesota Press, 1987.

Deswarte, Sylvie, and Bertrand Lemoine. *L'Architecture et les ingénieurs: Deux siècles de construction*. Paris: Editions du Moniteur, 1979.

Drew, Philip. *Frei Otto: Form & Structure*. Boulder, Colo.: Westview Press, 1976.

Ferguson, Eugene S. *Engineering and the Mind's Eye*. Cambridge, Mass.: MIT Press, 1992.

"Footbridge 2005: The Second International Conference, December 6–8, 2005." Venice, Italy: University IUAV, 2005.

Foug, Carolyn Ann, and Sharon L. Joyce, eds. "Reading Structures." *Perspecta 31, The Yale Architectural Journal*. Cambridge, Mass.: MIT Press, 2000.

Frampton, Kenneth. *Studies in Tectonic Culture: The Poetics of Construction in Nineteenth and Twentieth Century Architecture*. Cambridge, Mass.: MIT Press, 1995.

Francastel, Pierre. *Art & Technology in the Nineteenth and Twentieth Centuries*. New York: Zone Books, 2000.

Glaeser, Ludwig. *The Work of Frei Otto*. New York: Museum of Modern Art, 1972.

Giedion, Sigfried. *Space, Time and Architecture: The Growth of a New Tradition*. Cambridge, Mass.: Harvard University Press, 1941.

Gordon, J.E. *Structures: Or, Why Things Don't Fall Down*. New York: Plenum Press, 1978.

"Hybride Strukturen." *Archithese* 30, no. 3 (May-June 2000).

Jodard, Paul. "Technocracy: The Day of the Engineer." *World Architecture* 29 (1994): 72–75.

Kepes, Gyorgy, ed. *Structure in Art and in Science*. New York: George Braziller, 1965.

Khan, Yasmin Sabina. *Engineering Architecture: The Vision of Fazlur R. Khan.* New York: W. W. Norton & Company, 2004.

Kolarevic, Branko, ed. *Architecture in the Digital Age: Design and Manufacturing*. New York: Spon Press, 2003.

Le Corbusier. *Towards a New Architecture.* London: Architectural Press, 1927.

Leach, Neil, David Turnbull, and Chris Williams, eds. *Digital Tectonics*. Hoboken, N.J.: Wiley-Academy, 2004.

Lyall, Sutherland. *Masters of Structure*. London: Laurence King, 2002.

Lynn, Greg. "Blobs (Or Why Tectonics Is Square and Topology is Groovy)." *ANY: Architecture New York* 14 (1996): 58–61.

Mainstone, Rowland J. *Developments in Structural Form*. Oxford: Architectural Press, 1998.

Otto, Frei. *Tensile Structures: Design, Structure, and Calculation of Buildings of Cables, Nets, and Membranes*. Cambridge, Mass.: MIT Press, 1967–69.

Pacey, Arnold. *Meaning in Technology*. Cambridge, Mass.: MIT Press, 1999.

Peters, Tom F. *Transitions in Engineering: Guillaume Henri Dufour and the Early 19th Century Cable Suspension Bridges*. Basel: Birkhäuser, 1987.

Petroski, Henry. *To Engineer is Human: The Role of Failure in Successful Design*. New York: Vintage, 1992.

____. *Pushing the Limits: New Adventures in Engineering*. New York: Knopf, 2004.

____. *Success Through Failure: The Paradox of Design*. Princeton: Princeton University Press, 2006.

Pfammatter, Ulrich. *The Making of the Modern Architect and Engineer: The Origins and Development of a Scientific and Industrially Oriented Education*. Basel: Birkhäuser, 2000.

Picon, Antoine. "La Notion moderne de structure." *Les Cahiers de la recherche architecturale* 29 (1992): 101–10.

Picon, Antoine, and Alessandra Ponte, eds. *Architecture and the Sciences: Exchanging Metaphors*. New York: Princeton Architectural Press, 2003.

Rahim, Ali, ed. "Contemporary Techniques in Architecture." *Architectural Design* 72, no. 1 (January 2002): 4–96.

Rappaport, Nina. "Deep Decoration." In *Decoration: 306090.* Emily Abruzzo and Jonathan Solomon, eds. New York: Princeton Architectural Press, 2006.

____. "The Reception and Image of Modern Industrial Buildings." Proceedings of the Docomomo Eighth International Conference. (September 2003): 239–43.

Rickey, George. *Constructivism: Origins and Evolution*. Rev. ed. New York: George Braziller, 1995.

Riley, Terence, and Guy Nordenson, eds. *Tall Buildings*. New York: Museum of Modern Art, 2003.

Thompson, D'Arcy Wentworth. *On Growth and Form*. Cambridge: Cambridge University Press, 1961.

Torroja, Eduardo. *Philosophy of Structures*. Berkeley: University of California Press, 1958.

Tzonis, Alexander. *Santiago Calatrava: The Poetics of Movement*. New York: Universe Publishing, 1999.

ACKNOWLEDGMENTS

I am deeply grateful to all who assisted the process of making this book, especially the structural design engineers who supported me in my effort to understand the parameters and the depth of what Tim Macfarlane calls "their humble profession."

I am especially thankful to the Graham Foundation for Advanced Studies in the Fine Arts for a grant that enabled me to travel to meet the engineers and visit their projects and to the New York State Council on the Arts for an independent research grant.

I express sincere thanks to the staff of the engineers' offices especially Marc Ayala and William Dentzer, Arup; Cecilia Trollby, Atelier One; Susanne Nowak, Bollinger & Grohmann; Simon Kearney-Mitchell, Buro Happold; Prisca Schoenahl, Conzett Bronzini Gartmann; Susan Martinez, Dewhurst Macfarlane; Maggie Railton, Expedition Engineering; Derek Chan, Guy Nordenson Associates; Kinshasa Peterson, LERA; Anna-Inés Hennet, RFR; Hajime Narukawa and Isozaki Ayumi, Sasaki and Partners; Karin Hauber, Schlaich Bergermann und Partner; Dr. Frank Heinlein, Werner Sobek; and Chloe Sharratt, Jane Wernick Associates.

The encouragement of numerous colleagues assisted in the development of my research including Kenneth Frampton, whose last chapter of *Studies in Tectonic Culture* addresses the significance of engineers, James Axley, Frank Barkow, Inge Beckel, Deborah Gans, and Robert A. M. Stern.

Others over the years have been a stimulus to my work including Inge Beckel, Patrick Bellew, Lance Jay Brown, Annette Bögle, Keller Easterling, Ray Gastil, Jayne Merkel, Hilary Sample, Peter Cachola Schmal, Jonathan Solomon, Suzanne Stephens, Alexander Tzonis, Adam Yarinsky, and Katherine Dunne of Dunne and Markus engineers, who reviewed the book in its entirety as well as Kate Ellis, who provided me with both an office and support as a fellow writer.

Most importantly I would like to thank my assistants: Peter Arbour, Larissa Babiji, Marc Guberman, Maggie Hartnick, Amanda Webb, and especially research assistant Katherine Davies, and image editor Mathew Ford. Without them, this project would not have come to fruition.

At The Monacelli Press I would like to thank Gianfranco Monacelli, Andrea Monfried, Elizabeth White, and Stacee Lawrence. And my appreciation extends to Lara Stelmaszyk, a thoughtful and insightful editor; and to John Clifford of Think Studio who created the elegant design.

This project had its genesis in my work as a historian on the Historic American Engineering Record survey team of the Columbia River Highway in Oregon and visits to projects of Renzo Piano with engineer-architect Jean-François Blassel. But my interest in this topic could also be traced to something more innate. My grandfather, the late Leonard Shaffer, was a civil engineer in Philadelphia and I visited his structures as a child.

My warmest thanks go to my husband, Christopher Hall, whose investigative inquiry into the integration of structure and design led us to numerous Robert Maillart bridges in Switzerland. Without his keen editorial eye and inspiration, this book would not have been possible. My daughter, Alexandra, and my son, Adam, were stalwart traveling companions on multiple expeditions. And my parents were great enthusiasts.

CREDITS

Every effort has been made to acknowledge rights holders. We sincerely regret any errors or omissions.

© Peter Aaron / Esto (Courtesy of Apple): 98
© AFA / Arup: 22 bottom
Courtesy of Arata Isozaki: 171
Archimation, Berlin: 198 top
Archipress: 158 bottom
Courtesy of AREP: 164, 165 bottom right
Arup: 16–20, 23–25, 26 middle, 26 right, 28 right, 32
© Arup Cecil Balmond: 28 left
Arup / Markus Schulte: 30
Arup+CSCEC+PTW: 34–36, 37 bottom
Atelier One: 40 middle, 41, 44, 46 top, 47, 48, 49 bottom, 50
© Bollinger + Grohmann Ingénieure: 54, 56 top, 56 bottom, 57 top left, 57 bottom, 58 top, 59, 60 top, 61 top middle
© Cecil Balmond: 22 top, 26 left
Barkow Leibinger Architects: 90, 92 top right, 93
© Hélène Binet: 61 top right, 210 bottom
Marcus Bredt (Courtesy of von Gerkan, Marg and Partners): 184 bottom, 185–187
Richard Bryant/ARCAID: 94, 96 left, 112 bottom
© Marcus Buck: 61 top left
© Buro Happold: 68, 71 left, 72, 74 top, 76 top
© Friedrich Busam (Courtesy Bollinger and Grohmann): 56 middle
Conzett, Bronzini, Gartmann: 83 top left, 86–88, 89 top, 92 top left, 92 bottom
© Coop Himmelb(l)au: 58 bottom, 60 bottom, 61 bottom
Courtesy David Chipperfield Architects: 213
Wilfried Dechau: 80, 89 bottom
Dewhurst Macfarlane and Partners: 97, 100, 102, 104, 106 bottom
Gert Elsner / Schlaich Bergermann und Partner: 182 middle
Ernst-Haeckel-Haus, Friedrich-Schiller-Universität Jena, Germany: 11 right
H.G. Esch: 182 bottom, 201
Expedition Engineering: 110 bottom right, 111–119, 120 left middle, 120 right, 120 left bottom, 121
Chris Wise, Expedition Engineering: 108, 110 top left, 110 top right, 110 bottom left, 120 left top
Fake Design (China): 153 top
Brad Feinknopf (Courtesy of Rafael Viñoly Architects): 101, 103
© Mark Fisher: 42 top
Courtesy of Bernard Franken: 55
Courtesy Atelier Frei Otto: 11 bottom, 69 top, 70 bottom
Dennis Gilbert: 96 right
© John Gollings: 51
Mike Graham: 99 right
Guy Nordenson Associates: 138, 140 right, 141, 142, 145 top left, 145 top middle, 146, 148, 149 top left, 149 top right, 151, 152
Roland Halbe: 181 bottom, 206, 207
Christopher Hall: 12
Tim Hursley: 147, 149 bottom
Courtesy of Italcementi Archive: 140 left
Jane Wernick Associates: 210 middle left, 210 middle right, 211, 212, 214–218, 220
Niels Jonkhans: 62 bottom
© spacelab-Niels Jonkhans: 62 top
Sam Kittner: 135
Courtesy of Kohn Pedersen Fox Architects: 99 left
© Andreas Kunert/architekturphoto: 168, 169
Fundacion Le Corbusier/Artists' Rights Society: 6
Courtesy of Leslie E. Robertson Associates: 122, 124–132, 134
P. Marechaux © AKAA: 70 top
Peter Marlow (Courtesy of Atelier One): 40 top, 45
Courtesy of Massimiliano Fuksas Architetto: 181 top left
Arup / Ben McMillan: 37 top left, 37 top right
Michael Maltzan Architects: 150, 153 bottom
Matthias Michel: 64 bottom

Photography Collection, Miriam and Ira D. Wallach Division of Art, Prints and Photographs, The New York Public Library, Astor, Lenox and Tilden Foundations: 10 right
City of Chicago/ Walter Mitchell (Courtesy of Millenium Park): 40 bottom
Hidetaka Mori (Courtesy of The Esplanade Co. Ltd): 46 bottom
© Monika Nikolic: 63
Wilhelm Oester: 199
OMA/Ole Scheeren & Rem Koolhaas: 27, 29
© Paul Ott: 65
The Architectural Archives, University of Pennsylvania: 11 left
Polshek Partnership Architects: 133
Cedric Price Fonds. Collection Centre Canadien d'Architecture/Canadian Centre for Architecture, Montréal.: 9
Marc Quinn (Courtesy of Cass Sculpture Foundation): 42 bottom
Nina Rappaport: 8
RIBA Library Photographs Collection: 14
©Buro Happold/Mandy Reynolds: 71 right, 76 bottom, 77–79
RFR: 154, 156 left top, 156 left upper middle, 156 left bottom, 156 right, 157, 158 top, 158 middle, 159–163, 165 bottom left
Peter Rice / RFR: 156 left lower middle
Ralph Richter/Architecturphoto (Courtesy of Arup): 21
Christian Richters: 203, 205, 219, 221
John Riddy © Anish Kapoor (Courtesy of Baltic): 38
Tomas Riehle: 178
Marcus Robinson: 208
Andy Ryan (Courtesy of Arup): 31, 33
Andy Ryan: 136, 145 top right, 145 bottom
SANAA: 57 top right
Sasaki Structural Consultants: 166, 170, 172, 174, 176, 177
Schlaich Bergermann und Partner: 180, 181 top right, 182 top, 183, 184 top, 188–193
Simpson Gumpertz & Heger Inc: 144
© Margherita Spiluttini: 91
Squint Opera (Courtesy Grant Associates): 43
G. Stangl (Courtesy B+G Ingenieure): 64 top
© Dieu Tan: 49 top
D'Arcy Thompson, *On Growth and Form* (Cambridge University Press, first ed. 1912) From O. P. Hay, Iowa Geological Survey Annual Report, 1912.: 10 left
Courtesy of Toyo Ito & Associates, Architects: 173, 175
Dominique Uldry (Courtesy of Smarch): 84
Ruedi Walti: 82, 83 top right, 83 bottom
Ruedi Walti (Courtesy of Miller Maranta Architekten): 85
Paul Warchol: 139, 143
Werner Sobek: 194, 196, 197, 198 bottom, 200, 202, 204
Wilkinson Eyre: 105, 106 top, 107
© Buro Happold/Adam Wilson: 66, 69 bottom, 73–75
Zaha Hadid Architects: 210 top
© Gerald Zugmann: 52